AF594115

MAGNETIC COMPONENTS FOR POWER ELECTRONICS

MAGNETIC COMPONENTS FOR POWER ELECTRONICS

by

Alex Goldman
Ferrite Technology Worldwide, U.S.A.

KLUWER ACADEMIC PUBLISHERS
Boston / Dordrecht / London

Distributors for North, Central and South America:
Kluwer Academic Publishers
101 Philip Drive
Assinippi Park
Norwell, Massachusetts 02061 USA
Telephone (781) 871-6600
Fax (781) 681-9045
E-Mail < kluwer@wkap.com>

Distributors for all other countries:
Kluwer Academic Publishers Group
Distribution Centre
Post Office Box 322
3300 AH Dordrecht, THE NETHERLANDS
Telephone 31 78 6392 392
Fax 31 78 6546 474
E-Mail < services@wkap.nl>

Electronic Services < http://www.wkap.nl>

Library of Congress Cataloging-in-Publication Data

A C.I.P. Catalogue record for this book is available
from the Library of Congress.

Printed on acid-free paper.

Printed in the United States of America

TABLE OF CONTENTS

Preface

Power electronics is a rapidly-growing technology encompassing a large variety of applications including automotive, telecommunications, computers and alternative-energy systems. While the origin of modern power electronics was stimulated by the development of power semiconductor devices for high frequency switching purposes, other important and necessary accompanying component have been magnetic cores for transformers and inductors. This need has been especially true for the square wave features produced by solid-state switches that yield a plethora of higher harmonics. The heart of the new circuitry was the switched mode power supply (SMPS). As this new technology developed, there was an increased demand for miniaturization. Magnetic cores were traditionally the largest component in a solid-state circuit. Operation at higher frequencies was seen as a possible solution, As power semiconductors were designed for higher frequencies, the magnetic component suppliers were able to develop new materials and shapes that were optimized for the new conditions. Ferrites were found to be the most suitable materials for the low to medium wattage power supplies. In my first book, Modern Ferrite Technology, I said that "ferrites were the new kids on the block". While ferrites still maintain their dominance as power magnetic materials, "there are newer kids on the block ", namely amorphous and nanocrystalline materials. Although these materials are still in their infancy, they require more than a glancing look.

My purpose in writing this book was to review the many changes that have taken place in the past 5-10 years in power magnetic components. Some of these changes are;

1. Changes in the management of several of the major magnetic component suppliers. These suppliers included Siemens, Philips, Thomson and Allied- Signal (Metglas®).
2. Changes in the available power magnetic materials to operate at higher frequencies and higher DC bias.
3. Greater use of low-profile and planar core shapes
4. The use of Power Function Correction (PFC), Resonant Converter and Soft-Switching Circuits
5. Greater emphasis on EMI requirements in Power electronic cir cuits.
6. The development of cores using amorphous and nanocrystalline materials.

This book will discuss these changes and show how it affects power electronic component choice.

The main intent of this book is to advise power electronic design engineers and other magnetic component users as to the considerations that should be made in the choice of an optimum magnetic component for a particular application. This book was not intended to be a manual for the complete circuit design of a power electronic system. There are many other fine books that deal with this subject and some of the are listed in this book under the topic of design aids. Also missing from this book are the sections on basic magnetic theory and the physical, chemical and manufacturing aspects of ferrites. These are found in the author's book, Handbook of Modern Ferromagnetic Materials. There is, however, material-oriented information on the new amorphous and nanocrystalline materials since power components of these materials were not covered elsewhere. Aside from these new materials and ferrites, other metal strip materials and metal powder cores are also discussed. Suppliers of magnetic components may find this book useful as a guide to what properties are desired by the device or system engineer.

The first chapter deals with the various power electronic topologies including those involved with switched mode power supplies and resonant converters. The second chapter list the considerations encountered in the choice of a magnetic component including the material and shape. The third chapter expands on the material properties while the fourth chapter is involved with component shape including low-profile, planar cores and integrated magnetics. Chapter 5 is involved with the final choice of the size of the core and the windings needed. Included are some examples including the steps in determining the optimum size. These examples include applications for ferrites, amorphous cores and nanocrystalline cores. The sixth chapter contains a compendium of catalog data from the major magnetic component manufacturers in the World. While the data for each individual core are not included (the book would be voluminous), the material parameters and core shapes available are included. The final chapter includes aids for magnetic component design including books, articles, manufacturers CD-ROM's or diskettes and Web sites. Appendices include Tables of Units Conversion, Symbols, IEC and ASTM Standards and Addresses of Major Magnetic Component Suppliers.

Alex Goldman

ACKNOWLEDGEMENTS

I would like to thank Professor Robert W. Erickson for permission to use some material from his recent book, Fundamentals of Power Electronics, 2nd Edition. I would also like to thank Gordon "Ed" Bloom for permission to use some figures from his book, Modern DC-To-DC Switchmode Power Converter Circuits" and for discussions about this book. Thanks also are due to Colonel Wm. McLyman, a friend of many years for his useful design notes. Joe Huth III provided some photos of ferrite cores. I could not have written this book without the loving understanding of my wife, Adele, and the encouragement of my children, Mark, Beth and Karen.

Chapter 1

APPLICATIONS AND TOPOLOGIES FOR POWER ELECTRONIC SYSTEMS

INTRODUCTION-HISTORY OF POWER ELECTRONICS

Power electronics, as we know it, started in the late 1970's. There had been earlier use of power electronics but the development of power semiconductors was the impetus that made the field really take off. The nineteenth century was the century of D.C. power. At the turn of the 20th century, alternating current challenged D.C. with the debates of Edison and Tesla emphasizing the differences. Alternating current won out because power transformers allowed for better control of power and high voltage transmission and was more economical. Mechanical devices such as vibrators and rotary devices allowed inverters to produce ac from DC. Mercury-arc vacuum tube devices and commutators allowed the converse (producing D.C. from ac). During World War 2, magnetic amplifiers were developed which allowed better control of power. They, however, were costly and had lower efficiencies than expected.

In 1948, Bardeen, Brattain and Shockley invented the transistor but the early applications were for very low power applications such as in portable radios. The transistor was not used for power electronics to a large extent for about 25 years. Other semiconductor devices were developed during that time. High current semiconductor rectifiers, first using germanium and later silicon, replaced the mercury-arc rectifier. In 1957, a great advance in power semiconductor devices, the thyristor or SCR, was introduced. Between 1967-1977, the static VAR generator and some relatives of the thyristor (gate-controlled switch(GTO) and gate-assisted-turnoff device (GATT) were invented. At the end of the 70's, power transistors were sufficiently improved to challenge the thyristors for low power (several hundred watts) applications. In 1978, metal oxide semiconductor field-effect-transistors (MOSFET'S) challenged the power transistors and have become the staple of digital and analog devices. Other semiconductor devices for power electronics include the bipolar-junction transistor (BJT), the MOS controlled thyristor (MCT) and the insulated-gate-bipolar transistor (IGBT).

As the power semiconductor components were able to operate at higher and higher frequencies, the manufacturers of power magnetic components were challenged to develop the improved higher frequency materials to support the newer technologies. This has resulted in the change of materials for those that were used at the old 16 KHz. range to the present ones that are used in the MHz. range. In addition, the proliferation of printed circuit boards with power supplies mounted on them has led to a variety of new low-profile or planar power magnetic shapes.

1.1- APPLICATIONS FOR POWER ELECTRONICS

Power converters have two main functions common to most applications.

1. Power converters control the rate of power flow from a common supply to the load.
2. Often, power converters modify the power form to one more suitable for their loads.

The first is obvious. The second is somewhat more subtle. Electricity is rarely used per se (possibly only in electrochemical and metals industries). The users of energy have varying requirements as to the form of the electrical energy. Semiconductor power converters offer a means for supplying these diverse needs from a common source. They also do this with greater efficiencies than earlier methods. The broad categories for power electronics are;

1. Power Supplies-Computers and Microprocessors- These devices require a low-voltage DC bias for their operation. Semiconductor-based D.C. power supplies are more efficient, smaller in volume and lighter in weight than the linear variety.
2. Automotive- Applications in this area include car radios, electronic ignition, power steering, battery chargers and electric vehicle drive trains.
3. Telecommunications- 48 V. Distributed bus, front-end off-line power-factor correction(PFC), universal utility line.
4. Space Systems-Need for small size, low weight, efficient, reliable power source from batteries, solar and fuel cells.
5. Motor Controls-Speed and Torque control systems, D.C. motor control, disk drives
6. Lighting-Electronic- fluorescent-light ballasts,
7. Alternative Energy-Conversion of power from wind, water and sunlight to a useable form.
8. Un-interruptible Power Supplies (UPS)

1.2- POWER CONVERTERS

The heart of the power electronics system is the power converter. The term, power controller, can have several different meanings. To an electrical engineer, it may refer to one of several different basic combinations and circuits. These include the Buck, Boost and Buck-Boost Converters. To a magnetics engineer, it may refer to one of the most commonly-used transformer-coupled circuits made up of variations of the above basic circuits. These circuits include the Forward, Flyback and Push-Pull Converters. Power Converters can also be classified according to their function. These include;

1. DC-DC Converters
2. ac to ac converters
3. DC to ac converters
4. ac to DC converters

The last of these converters is a rectifier while the third is an inverter. The layout of a converter circuit is shown in Figure 1.1.

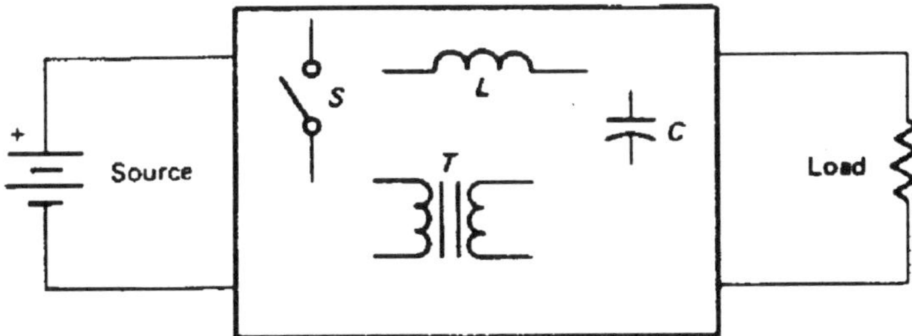

Figure 1.1- Schematic of a Power Converter, From Severns and Bloom, Modern DC-to DC Switch-Mode Converter Circuits, (1984) EJBloom Associates Inc., San Rafael CA

The unusual feature contained in these components is the absence of resistors which were used in earlier power conditioners and which produce high losses. Thus, the efficiencies of the new power electronic systems are much greater. The basic power converter includes an input, output and an intermediate converter circuit composed of switches, transformers, magnetic components and capacitors. Ideally, most of these components are without the high losses of resistors. The magnetic components are transformers and chokes while the semiconductor switches are linear BJT(Bipolar Junction Transistor) and IGBT (Insulated Gate Bipolar Transistor) or Switched-Mode MOSFET(Metal Oxide Semiconductor Field Effect Transistor)/ These components are shown in Figure 1.2. The manner in which these elements are combined is called the topology. The basic topologies are described in the next section.

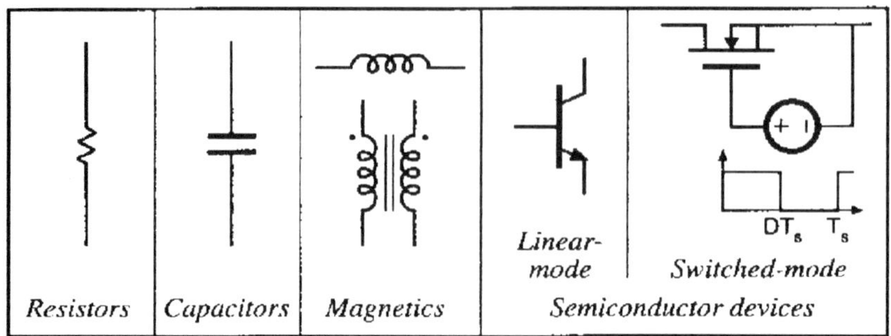

Figure 1.2- Components used in a Converter Circuit. From Erickson, Fundamentals of Power Electronics, Second Edition, Kluwer Academic Publishers, Boston (2001)p.3

1.3-BASIC TOPOLOGIES OF POWER CONVERTERS

The simplest DC circuit containing a switch is one that only has a load resistance in series with the switch. This circuit is shown in Figure 1.3a. with the voltage wave form shown in Figure 1.3b. When the switch is in the 1 position, the output voltage is V_g and when the switch is in the 2 position the output voltage is 0. A rectangular wave-form is produced as shown in Figure 1.3b.

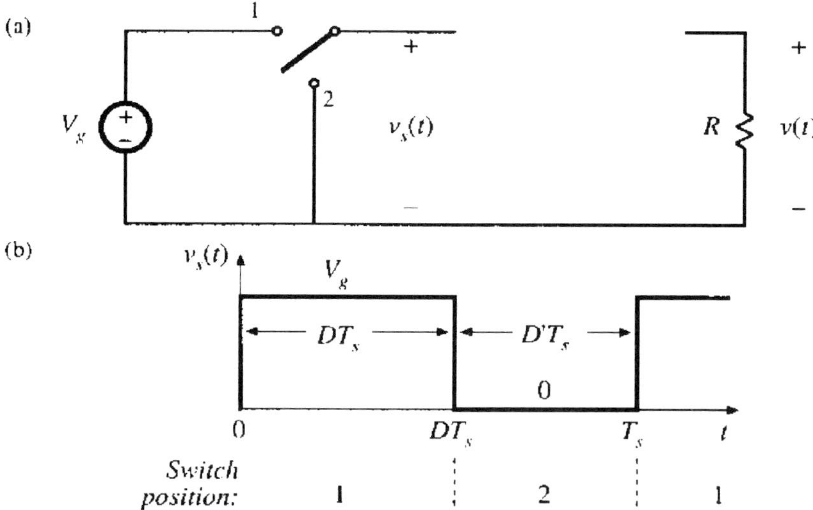

Figure 1.3- a) A Circuit containing an input, a switch, and a load resistance. b) The output wave-form when the switch is moved from the 1 to 2 position From Erickson, Fundamentals of Power Electronics, Second Edition (2001) p.13

The average output voltage over the switching period, T_s , is the integral of the voltage with respect to time or the shaded area in the curve as shown in Figure 1.3c. This integral can be reduced to DT_s where D is the duty cycle or the ratio of time on over total switching cycle time. Thus the voltage can be regulated from 0 to V_g by variation of the duty cycle. If a low-pass filter is added to the circuit, the DC will be passed while eliminating the ac switching frequency. One such topology is the Buck converter circuit. Here, the low-pass filter is added after the switch. A schematic of this circuit is shown in Figure 1.4

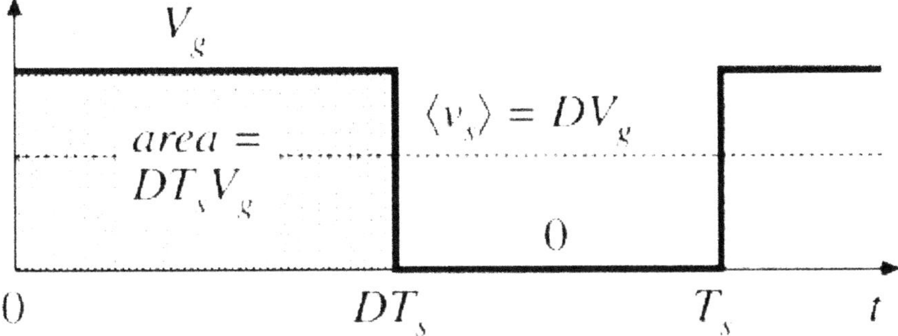

Figure 1.3c-Determination of the switch output voltage DC component by integrating and dividing by the switching period From Erickson, Fundamentals of Power Electronics, Second Edition (2001) p.14

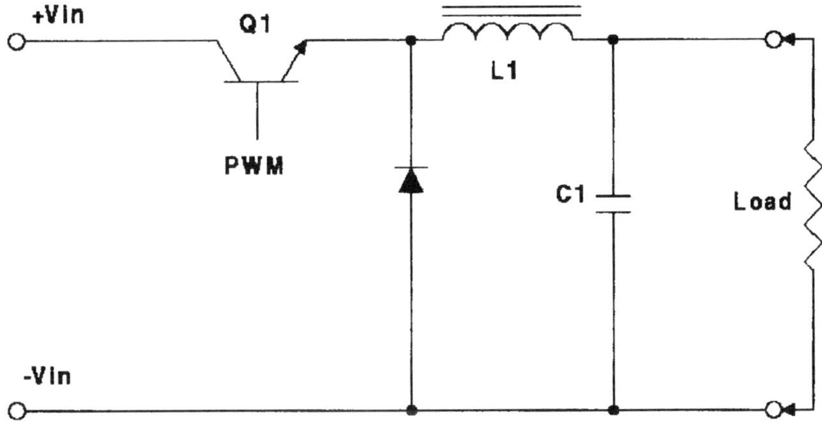

Figure 1.4-A Schematic of a Buck Converter

Similar to the case of Figure 1.2 with only the resistor in the circuit, the output voltage is given by DT_s (Figure 1.5) in which D<1, hence the name Buck or step-down converter.

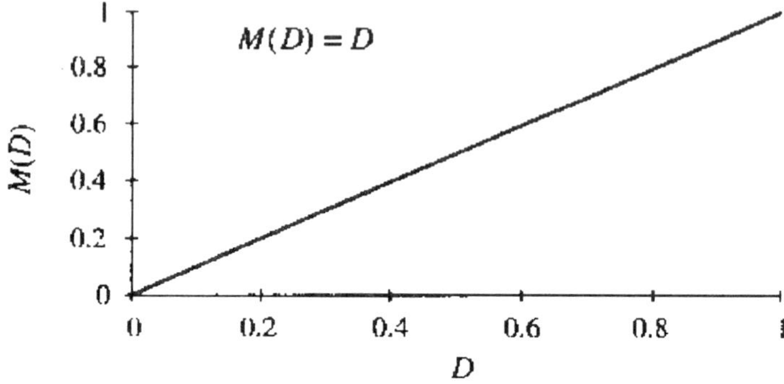

Figure 1.5-Variation of output to input voltage ratio with duty cycle for a Buck converter From Erickson, Fundamentals of Power Electronics, Second Edition (2001) p.16

If the switch and inductor are exchanged another converter topology is produced. In this case the output voltage is increased. This Boost or step-up converter type is shown in Figure 1.6

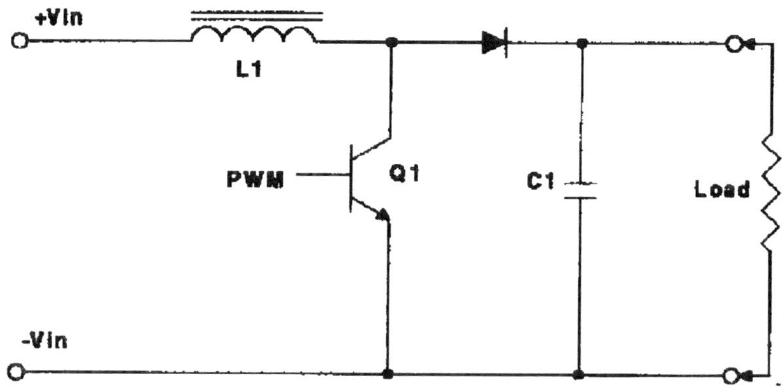

Figure 1.6 -**A Schematic of a Boost Converter**

The variation of conversion ratio, V/V_g with duty cycle for a Boost Converter is shown in Figure 1.7

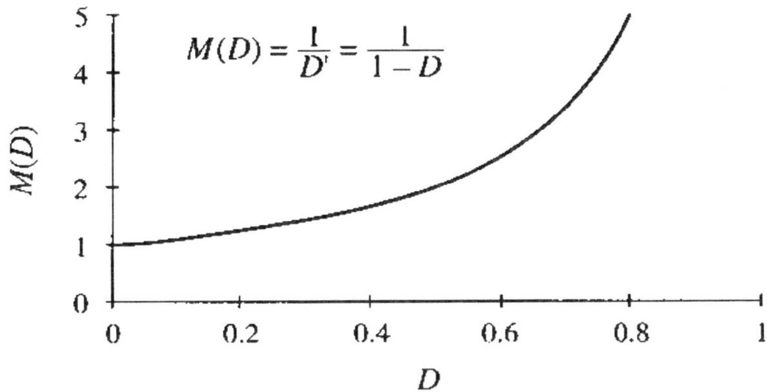

Figure 1.7-Conversion Ratio of a Boost Converter as a function of Duty Cycle From Erickson, Fundamentals of Power Electronics, Second Edition (2001) p.16

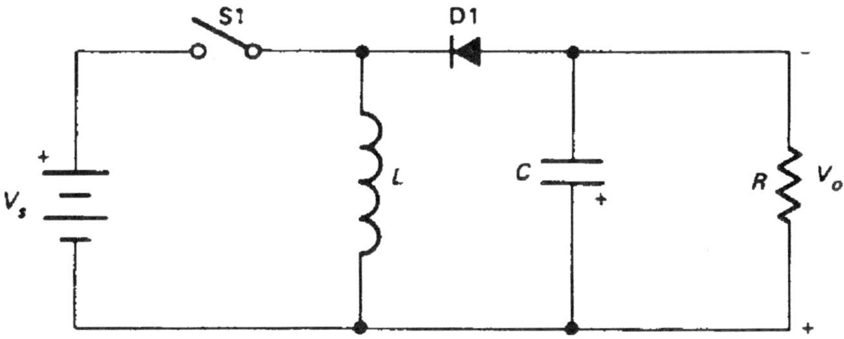

Figure 1.8-Schematic of a Buck-Boost Converter

By combining several of the simple buck or boost converter circuits many other useful and popular converter topologies can be formed. One such example is the buck-boost converter. The buck-boost converter has the same number of elements as the buck or boost converters but is considered a cascading of the two types with the elimination of one of the inductors. The conversion ratio is also D/(1-D) allowing high voltage ratios. However, this topology cannot be use in high power applications. A schematic of a buck-boost converter is shown in Figure 1.8

1.4-PRACTICAL CONVERTER TOPOLOGIES

The converter circuits described in previous sections were simple ones without the use of transformers between the input and output. The use of such a transformer accomplishes three main objectives;

1. D.C. Isolation- Isolation is important for both safety and circuit performance concerns.
2. Where large transformation ratios are desired, addition of a transformer is more effective than with the previous simple designs.
3. The use of a transformer permits multiple outputs.

When transformer isolation is included in power supply design, several practical topologies with regard to the magnetic components are used. They are derived from the basic topologies described earlier with some variation along with the use of the transformer. The three most popular of these are;

1. Flyback Converter
2. Forward Converter
3. Push-Pull Converter

The type of converter circuit used will influence the operating conditions and, therefore, the choice of magnetic core to be used.

1.4.1- Flyback Converter

The flyback is one of the simplest designs for a converter. It is based on the book-boost design with the inductor replaced by a transformer. When the transistor switch is in the closed position, the transistor is in the conducting mode and the input voltage appears across the inductor, creating a magnetic field that stores the magnetic energy until the switch is opened. At that time, the current across the inductor reverses as the magnetic flux in the core decreases. The stored magnetic energy is transferred to the plates of the capacitor and the load. One way of varying the amount of energy stored and therefore the output voltage is by varying the **ON** time or duty cycle, D, of the transistor. The simplicity of design and fact that no output choke is needed make it a useful choice of design. Because it is unidirectional, a larger core may be needed than that used in other designs. It also has a higher "ripple" or residual ac component. A block diagram a flyback converter with multiple outputs is shown in Figure 1.9.

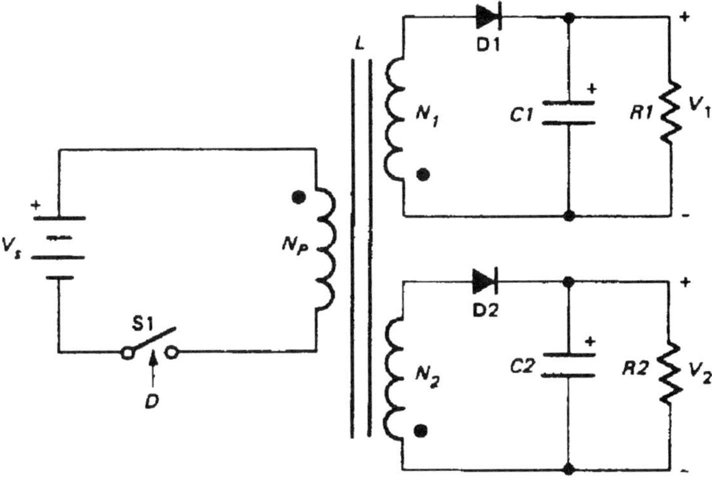

Figure 1. 9.-A schematic of a Flyback Converter with multiple outputs. From Severns and Bloom, Modern DC-to-DC Power Converter Circuits (1985) p.160

1.4.2-Forward Converter

Another type of converter, known as the forward converter is shown in Figure 1.10 (Bracke 1983). When the switch is closed and the transistor conducts, current again rises linearly in the inductor. However, since the inductor and load are in parallel to each other, part of the energy is stored in the inductor and part is transferred to the load. When the switch is opened, the current continues to flow to the load by way of the diode, known as the flywheel diode. The voltage can again be controlled by pulse width modulation. The forward converter is sometimes called the series converter and the flyback the parallel converter. The forward converter has lower ripple than the flyback as the combination of capacitor and inductor provides a better filter circuit. A block diagram for a forward converter is shown in Figure 1.10

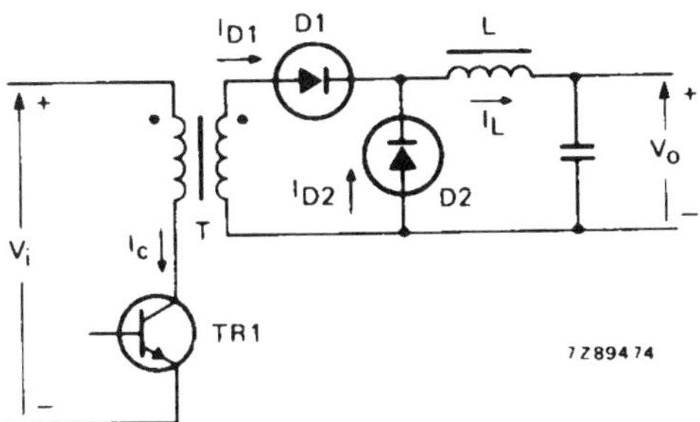

Figure 1.10-Forward Converter with transformer isolation. From Bracke and Geerlings(1982)

1.4.3. Push Pull Converters

The push-pull converter is essentially a combination of two forward converters operating in opposite directions, giving it the name push-pull. It has two separate transistor switches, one for each forward converter with a split center-tapped winding. When S1(TR1) is open, diode D2 conducts and energy is partially stored in the inductor and partially supplied to the load. With both S1 and S2 (TR1 and TR2) open, the inductor will provide a load current by parallel diodes, D1 and D2. When S2(TR2) closes, D1 will still conduct and D2 will stop and the process repeats itself. A push-pull converter has twice the ripple frequency and therefore a lower ripple-voltage. Another advantage is its ΔB swing is twice as large as the unipolar variety and therefore, a smaller transformer can be used. Multiple outputs can be constructed using several secondary windings. A block diagram is shown in Figure 1.11

1.5-CONVERTER STRENGTHS AND WEAKNESSES

Flyback transformers are really more like power inductors than the other types of transformers since the energy is stored in the inductance during the current rise and discharged during the flyback period. The design, therefore, will be considered under the subject of power inductors. Variations of

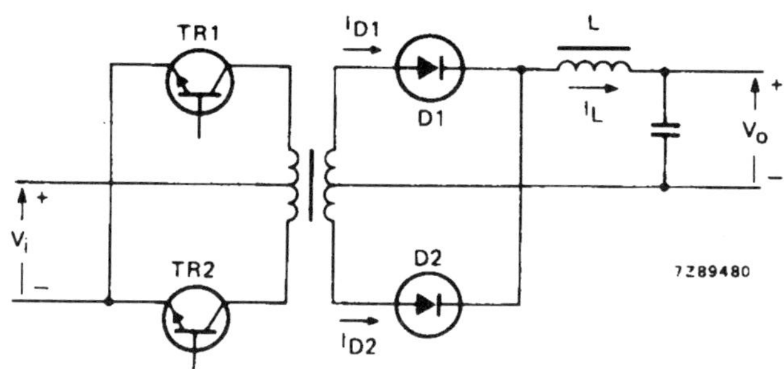

Figure 1.11-A push-pull converter From Ferroxcube Design Manual

these circuits are possible. Table 1-1 gives the relative strengths and weaknesses of each type. For high power applications, the push-pull design is preferred. For less demanding versions, the forward converter is an alternative. In high voltage supplies, the flyback type is most suitable.

In the design of transformers for inverters, the worst case scenario is used with regard to transient voltages that may increase the input voltage. Knowing the maximum and minimum voltages will help in the design process. Another

Table 1.1
Circuit Type Summary

Circuit	Advantages	Disadvantages
Push-pull	Medium to high power	More components
	Efficient core use	
	Ripple and noise low	
Feed forward	Medium power	Core use inefficient
	Low cost	
	Ripple and noise low	
Flyback	Lowest cost	Ripple and noise high
	Few components	Regulation poor
		Output power limited (<100 Watts)

operational problem that must be considered in the design of push-pull converters is the possibility of D.C. imbalance in the two arms of the circuit. For this reason, full bridge converters are used for most high-power applications even though they have twice as many power semiconductor switches. The voltage stress is only on the DC Bus voltage, not twice the value. In addition, a DC blocking capacitor can be added to the full bridge where it cannot in a push-pull circuit.

1.6-THE HYSTERESIS LOOPS FOR POWER MATERIALS

Since this book is involved with the application and choice of magnetic components in power electronics, it is important to relate the action and properties of the magnetic materials in the transformers and inductors in the circuits we have been studying. What produces the voltage is not the alternation but the rate of change of the flux. As the current and voltage wave-forms change during operation, the magnetic components go through the hysteresis loop characteristic of the magnetic material. In some cases only part of the hysteresis loop is traversed. The operation of a power transformer or choke can be designed to have a bipolar drive as in the push-pull type or unipolar (forward or flyback mode). In the bipolar case, the course of the induction or the excursion is in both directions so that the magnetization is reversed. The flux density or induction, B, traversals for the case for push-pull (a), forward (b) and flyback (c) topologies are given in Figure 1.12.

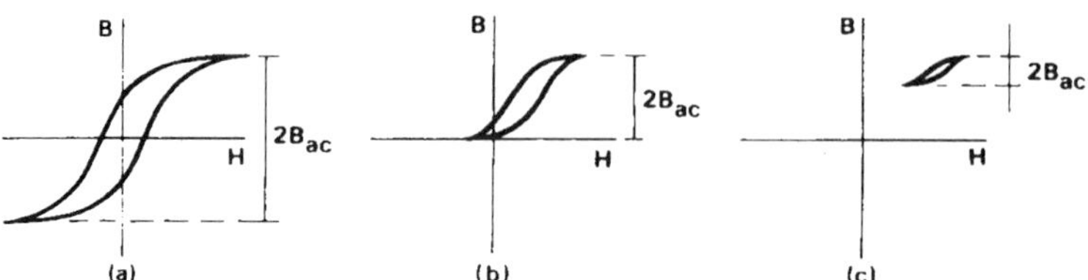

Figure 1.12- Flux Density or Induction traversals for (a) Push-Pull, (b) Forward and (c) Flyback Converters. From Bracke (1983)

In the unipolar case, the induction is unidirectional and the magnetization is not reversed. In Figure 1.12b, for the forward converter, a slow-rise capacitor or ringing choke has been added to reset the core. In the case of the flyback converter, (Figure 1.12c), the offset of the ac induction loop is due to the DC usually present in flyback converter. In certain instances in power electronics, the limits of induction are from the remanent to a higher induction. The loop traversed is a minor loop in the first quadrant similar to the one shown in Figure 1.13. In this case, the B used in the induction equation is still ΔB/2 even though the magnetization is not reversed.

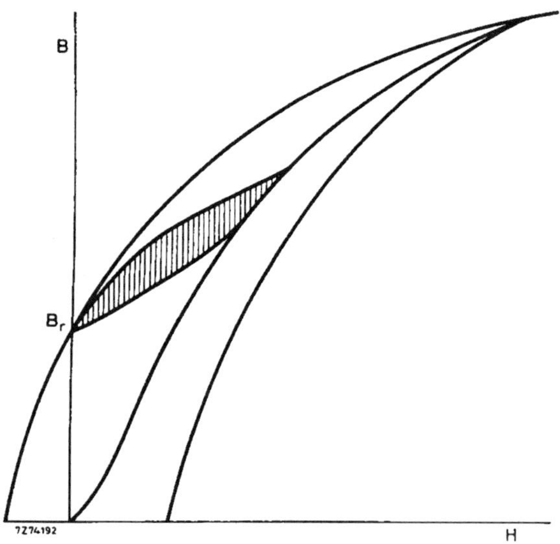

Figure 1.13-Minor hysteresis loop traversed when unipolar pulses are applied to a feed-through converter. The hatched area represents the portion of the hysteresis loop actually used. From Ferroxcube Application Note F602

Whereas the ΔB and thus the corresponding voltage are smaller in a unipolar than in the bipolar drive, the construction and operation are much simpler and more economical. The hysteresis loop traversals for flyback, forward and push-pull converters are shown in Figures 1.14, 1.15, and 1.16 . In the use of ferrites as flyback transformers for television receivers, the voltage and current waveforms are not sinusoidal but saw-tooth or flyback shape. This permits the electron beam to sweep across the TV screen with its visual signal and when it comes to the end, it would rapidly return or fly back to the starting horizontal sweep position to present the next lower line of information. In this case, the transformer operation is unipolar. The basis of the modern electronic switching power supply is the action of the transistor as a switch. Early transistors were not built to carry much power and thus, as was the case for early ferrite inductors and transformers, they were used mainly in telecommunication applications at low power levels. The first power electronic component was the inverter that is a device that takes a DC input and produces an ac output in a manner other than the usual rotary generator. A transformer may be incorporated in the device to give the required voltage. The device can be mechanical such as a vibrator or chopper or it can be of the solid state variety using a transistor. The word, oscillator, may sometimes be confused for an inverter but in the oscillator, the frequencies may be higher and the power levels lower. The second important item, a DC converter, takes the DC of one voltage and converts it to DC at another voltage. One might call it a DC transformer. The intermediate step in a converter is that of an inverter namely the conversion of DC to AC. Of course, the additional step is rectification to D.C. The input to a converter can sometimes be a low frequency (50-60 Hz.) which is rectified, inverted, transformed, and then again rectified. The advantage over a conventional transformer is that the transformation is much more efficient at the higher frequency.

1.7-SWITCHING POWER SUPPLIES

The complete switching power supply may consist of several auxiliary sections in addition to the power transformer. If ac is the input, it must first go through a noise filter to keep out unwanted transients. It is then rectified before entering the power transformer where it is first inverted to a square wave of the higher frequency (or pulse repetition rate) and then transformed to the desired output voltage. The transistor is driven by an auxiliary timing transformer or driver. After passing through the power transformer, the secondary voltage is again rectified. It then passes through a voltage regulator to maintain the voltage limits in the required range. Often this is done in a feedback circuit which controls the on-off ratio of the switching transistor. This

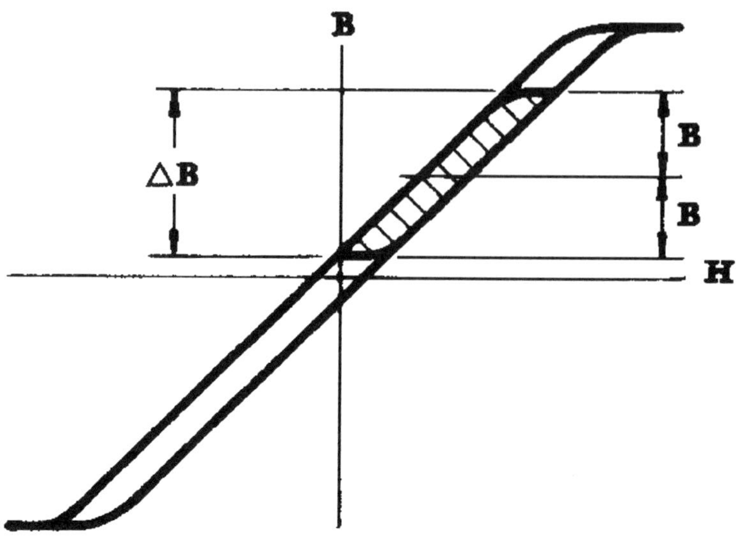

Figure 1.14-The hysteresis loop traversal during operation of a flyback Converter. From W.A. Martin, Proc., Power Electronics Conf.(1987)

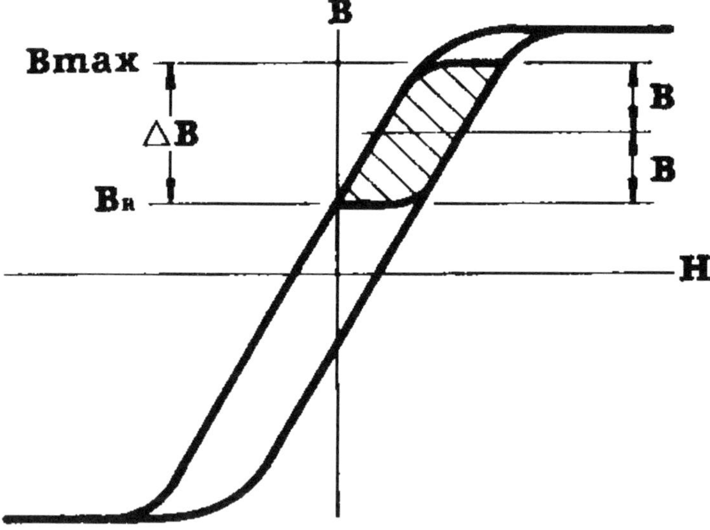

Figure 1.15- The hysteresis loop traversal during operation of a Forward Converter. From Martin (1987)

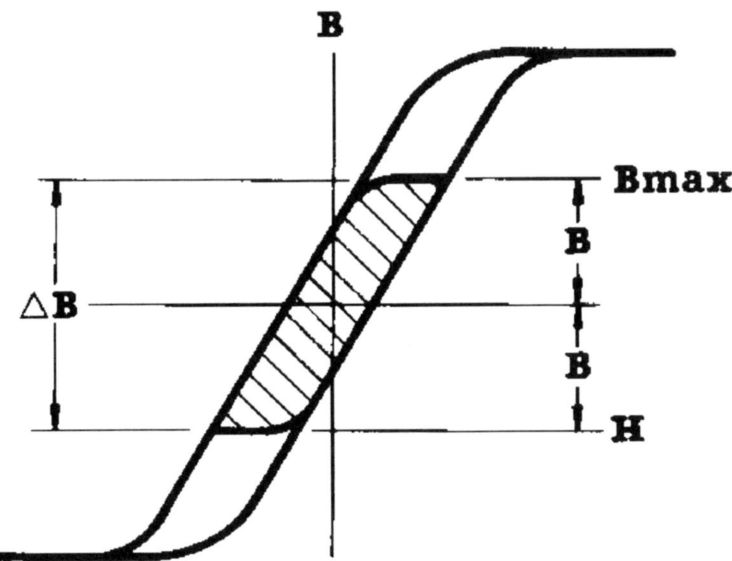

Figure 1.16- The Hysteresis Loop Traversal during Operation of a Push-pull Converter From Martin(1987)

technique is called "pulse width modulation" or PWM and is widely used. An example of such a circuit is shown in Figure 1.17. Switching power supplies have efficiencies on the order of 80-90% compared to those of linear power supplies that may range from 30-50%. The switching supplies are, therefore, lighter and smaller than their counterparts. A typical switching power supply is shown schematically in Figure 1.18 (Magnetics 1984).

1.8-FERRORESONANT CONVERTERS

In a previous section, we have discussed the use of pulse width modulation (PWM) as a means of regulating a high frequency power supply. This method using square wave produces high switching loss because of all the odd harmonics produced by the square wave. There are several other control mechanisms which we will discuss One is the use of a resonant or ferroresonant converter. The other is the use of a magnetic amplifier.

There are many instances of resonance as it relates to low level linear ferrite components. In such cases a series or parallel combination of an inductor and capacitor acted as an LC circuit for frequency control in low level filters. The term resonance (more properly, ferroresonance) here has more of a

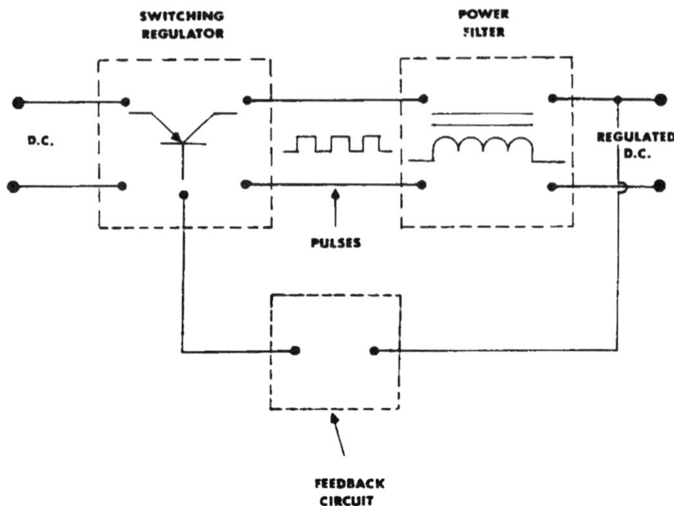

Figure 1.17-An example of a switching regulator using feedback by pulse width modulation to control the output voltage. From Magnetics PS-01(1982)

connotation of resistance to changes in the input voltage and current by storing energy in the resonant circuit. As a matter of fact the first uses of ferroresonance was in the construction of a constant-voltage 60 Hz. transformer by Sola. In power supplies, an important use of the ferroresonant transformer is as a regulator. The early 60Hz transformers have given rise to the high-frequency type, which as noted earlier, may be even more useful at the highest frequencies than the conventional switching transformer design.

As a high-frequency power inductor, the ferroresonant transformer has a quite different function . For one thing, the magnetic circuit is non-linear and because of the high currents and fields, operation is close to saturation. Most often when used as a power inductor, it is necessary to insert an air gap or spacer to avoid saturation.

Figure 1.19 shows a simple ferroresonant regulator that consists of a linear inductor, L_1, a non-linear saturating inductor, L_2 and a capacitor, C_1 in parallel with L_2. The latter two components form the ferroresonant circuit that controls the input voltage. The input energy is stored in L_1 and the resonant circuit acts to pass a uniform voltage to the load. Although the linear transformer may be of the typical power ferrite found in transformers, the saturating transformer is quite different. In addition to the usual attributes of power ferrites, it should possess a rather square hysteresis loop. The squareness ratio, B_r/B_s should be over 85%. The permeability over the linear portion of the loop

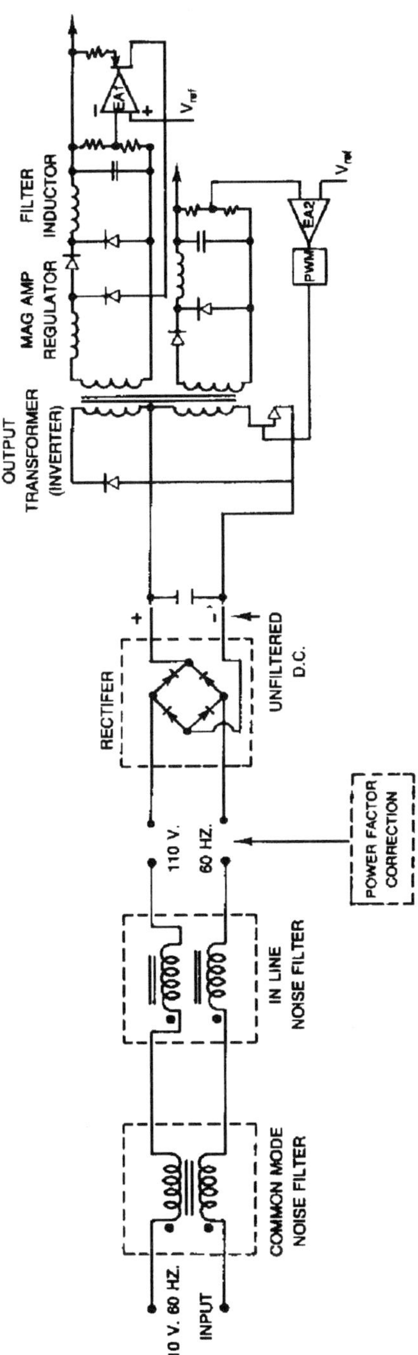

Figure 1.18.-Block diagram of a switching power supply From Magnetics PC-01(1982)

should be as high as possible with the saturation permeability quite low (μ = 20-30)

By combining the ferroresonant regulator with a high frequency inverter, a ferroresonant converter can be constructed as shown in Figure 1.20 Then, with the addition of a rectifier in front, a switching power supply can be made. See Figure 1.21.

McLyman(1969) has shown how a high frequency ferroresonant transformer, tuned to about 20 KHz. can be used to stabilize high frequency inverters.

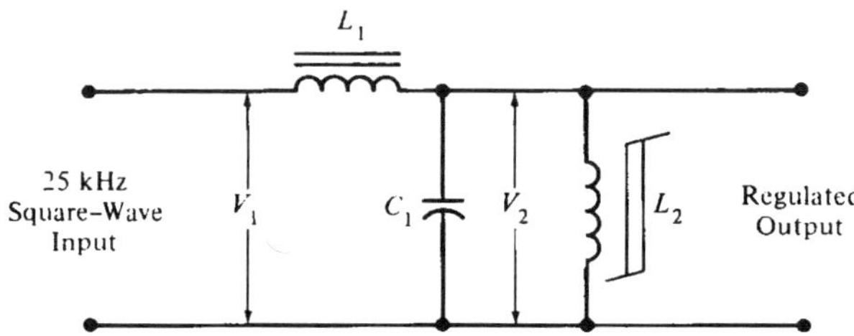

Figure 1.19-A simple ferroresonant inverter consisting of a linear inductor L1 and a non-linear saturating inductor

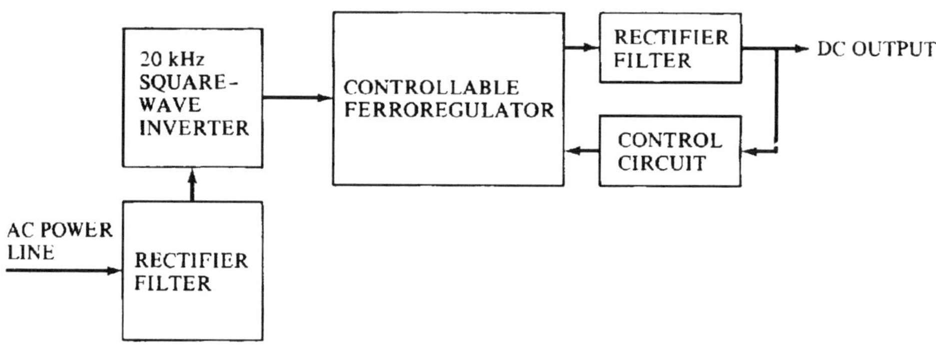

Figure 1.20- A Ferroresonant inverter. From Fletcher (1977)

With regard to the ac component, the permeability of the gapped core is larger than one operating at saturation. The amount of gap depends on the maximum D.C. current, the shape and size of the core and the inductance needed for energy storage. The a.c. ripple is usually on the order of 10% of the D.C. signal.

To estimate the maximum current, I_m, an extra 10% or more safety factor for transients is inserted in the design making I_m on the order of 1.2-1.3 I_0.

There are applications in which a high-frequency inverters is needed with low harmonic distortion. In a resonant inverter, a square-wave is produced by the switch network. The square wave has many odd harmonics that produces distortion and high switching loss. When the square wave is fed into the resonant tank circuit that is tuned to the fundamental frequency, a pure sine wave is produced that is then rectified and filtered. By changing the frequency closer or further from the resonant frequency, the voltage and current at the load can be controlled. A resonant inverter circuit with two switches is shown in Figure 1.22.With the addition of a rectifier and low pass filter network, a resonant converter circuit is formed. See Figure 1.23.

1.9-SOFT SWITCHING IN COMMON TOPOLOGIES

The reduced switching loss is the chief advantage of a resonant converter brought about by either *zero-current switching* (ZCS) or *zero-voltage switching* (ZVS). These fall under the heading of *soft switching* that can also

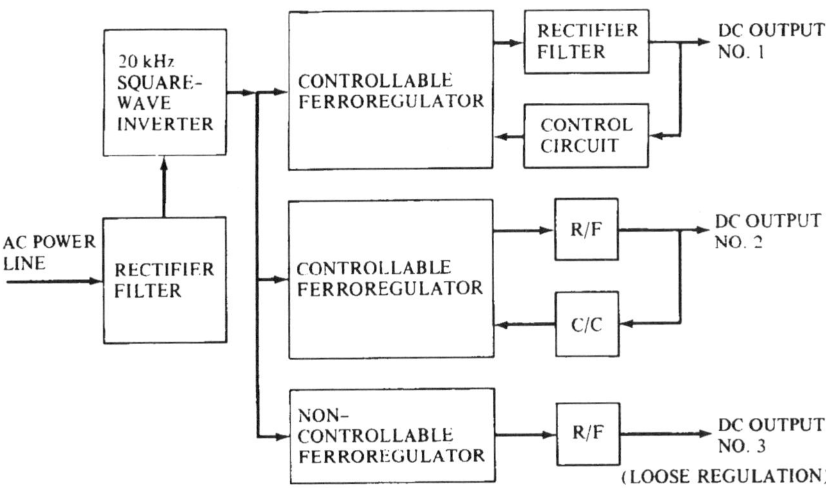

Figure 1.21- A Ferroresonant switching power supply From Fletcher (1977)

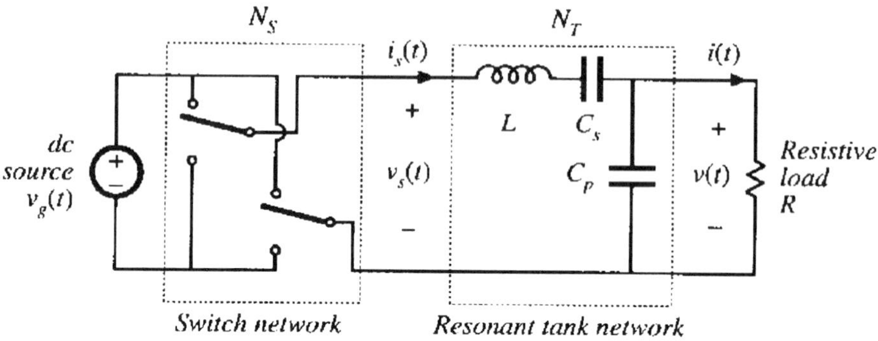

Figure 1.22-A Resonant Inverter Circuit From Erickson, From Erickson, Fundamentals of Power Electronics, Second Edition (2001) p.706

be applied to other topologies. The switching using the resonant inverter is done at the zero crossing points of the sinusoidal current or voltage waveforms and thus reduce the semiconductor switching loss allowing operation at higher frequencies. In a buck converter, zero-current switching can be implemented by insertion of a quasi-resonant switch cell in place of the PWM switch cell as shown in Figure 1.24 . It is also possible to insert a ZVS switch cell into a buck converter as shown in Figure 1.25. A Zero Voltage Switching Circuit using an active-clamp snubber network in a forward and flyback converter is shown in Figure 1.2.

SUMMARY

This chapter has reviewed the various circuits that are commonly used in power electronics. The hysteresis loop traversals of the magnetic components were correlated with the three topologies for unipolar and bipolar cases. The magnetic functions of the magnetic components involved in the operation a switching power supply will be discussed in Chapter 2.

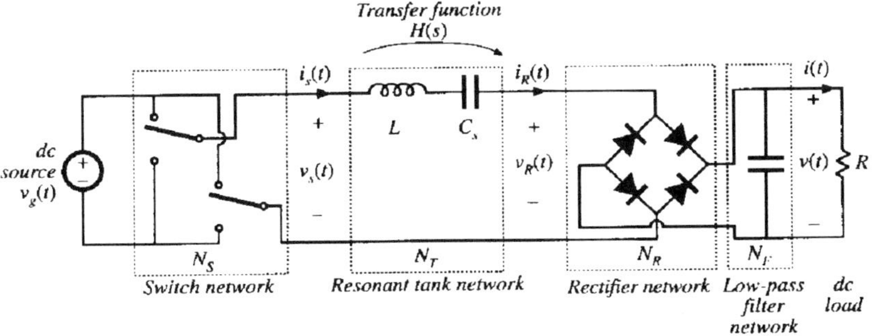

Figure 1.23-A resonant converter based on a resonant inverter, rectifier and filter network. From Erickson, Fundamentals of Power Electronics, Second Edition (2001) p.707

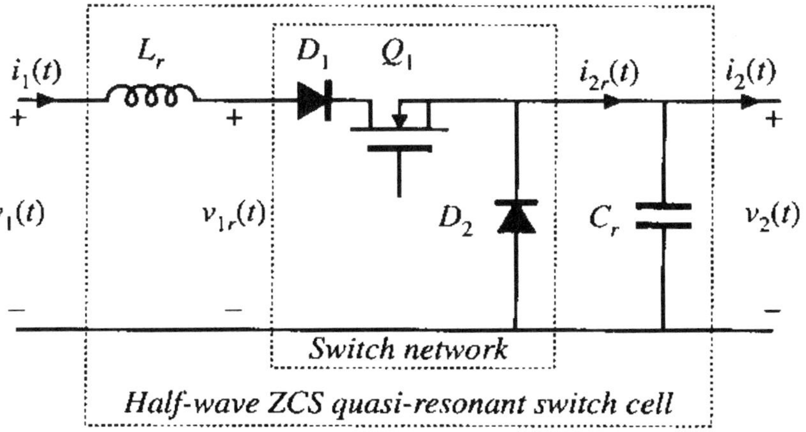

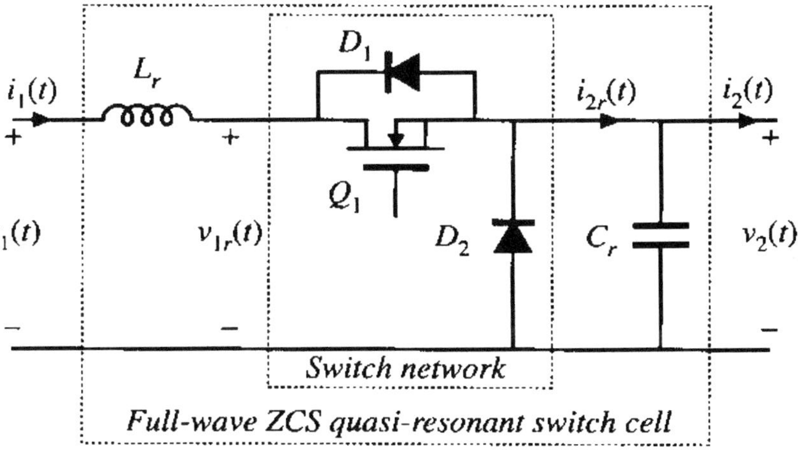

$i_1(t)$ L_r D_1 Q_1 $i_{2r}(t)$ $i_2(t)$

$v_1(t)$ $v_{1r}(t)$ D_2 C_r $v_2(t)$

Switch network

Half-wave ZCS quasi-resonant switch cell

D_1 $i_1(t)$ L_r $i_{2r}(t)$ $i_2(t)$

Q_1 $v_1(t)$ $v_{1r}(t)$ D_2 C_r $v_2(t)$

Switch network

Full-wave ZCS quasi-resonant switch cell

Figure 1.24 -A Half-Wave and Full-Wave Resonant Switch Cell for ZCS. From Erickson, Fundamentals of Power Electronics, Second Edition (2001) p.769

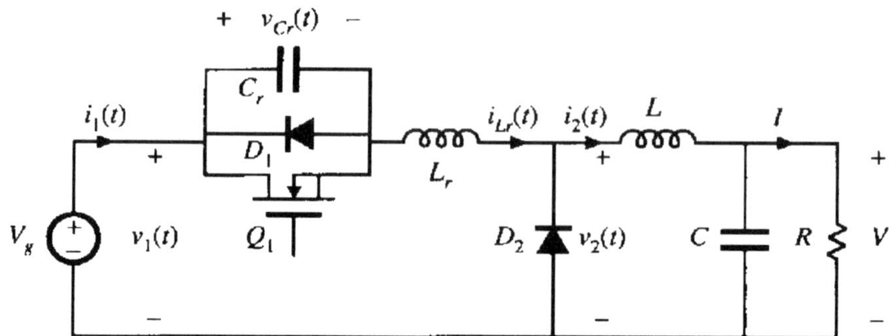

Figure 1.25-A ZVS Switch Cell in a buck converter. From Erickson, Fundamentals of Power Electronics, Second Edition (2001) p.784

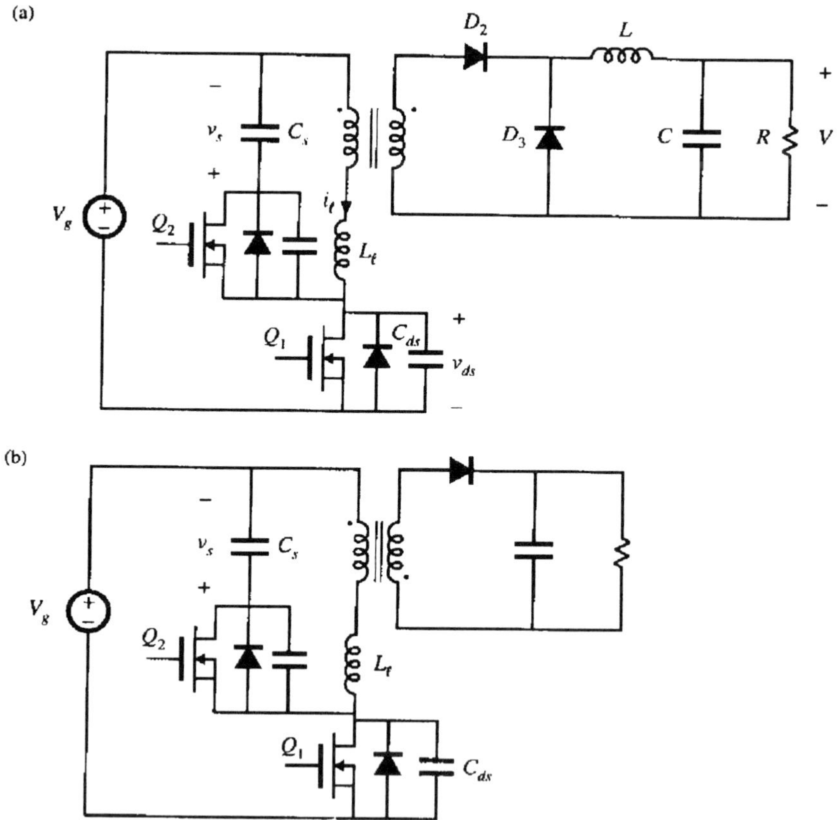

Figure 1.26-Zero Voltage Switching using an active-clamp snubber network in a forward and flyback converter. From Erickson, Fundamentals of Power Electronics, Second Edition (2001) p.795

References

Severns, R. P. and Bloom, G. E.,(1984) Modern DC-to-DC Switchmode Converter Circuits, EJ Bloom Associates, San Rafael, CA

Bracke, L.P.M.,(1983) Electronic Components and Applications, Vol.5, #3 June 1983,p171

Bracke L.P.M.(1982) and Geerlings, F.C., High Frequency Power Transformer and Choke Design, Part 1, NV Philips Gloeilampenfabrieken, Eindhoven, Netherlands

Erickson, R.W. and Maximovic, D., (2001)Fundamentals of Power Electronics, Second Edition Kluwer Academic Publishers, Boston, Dordrecht

Magnetics (2000) Ferrite Core Catalog, Magnetic Div.,Spang and Co, Butler, PA 16001

Martin, W.A.(1987), Proceedings, Power Electronics Conference (1987)

Chapter 2

MAIN CONSIDERATIONS FOR MAGNETIC COMPONENT CHOICE

INTRODUCTION

After having reviewed the applications and topologies for the various power electronic circuits in Chapter 1, we can now proceed to investigate the main considerations that are made in choosing the appropriate magnetic component. In this chapter, these considerations are related to the function of the component in the circuit. They include a general review of material properties and the core shapes and sizes. The final size will be determined by the design considerations and the input and output variables. Lastly, the question of cost will come into play depending on the market that the component and the finished product is aimed.

2.1-CONSIDERATIONS BASED ON COMPONENT FUNCTION

At the end of Chapter 1, we listed the various magnetic functions that are used in a switching power supply. They are;

1. Power Transformer
2. Power Inductor or Choke
3. In-line or Differential-Mode Choke
4. Common-mode Choke
5. EMI Suppression Core
6. Pulse Transformer for Transistor Firing
7. Magnetic Amplifier Core
8. Power Factor Correction Core

In choosing the best magnetic component for these functions, we use a process similar to that employed for low-level components except that the necessary parameters are somewhat different. The choice will be determined by:

1. The type of converter circuit used.
2. Frequency of the circuit.

3. Power requirements.
4. The regulation needed (percentage variation of output voltage permitted)
5. Cost of the component.
6. Efficiency required.
7. Input and output voltages

The components for each of these functions will be considered separately as their requirements are somewhat different. The cores selected to meet these requirements will be discussed with regard to;

1. Core material
2. Core configuration which includes associated hardware (bobbins, clamps, surface-mount connections etc.)
3. Size of the core
4. Winding Parameters- (number of turns, wire size)

This chapter will review these requirements in general for the various functions and in much more detail in the following chapters.

2.2-MAGNETIC COMPONENT CHOICES

Having limited the choice of material somewhat to only power and power-related applications, there are still a large variety of components that can be used. These include;

1. Soft ferrite cores
2. Powdered metal toroids (and some E-cores)
3. Magnetic metal strip cores (Tape cores, laminations, cut-cores)
4. Amorphous metal strip wound tape-cores
5. Nanocrystalline material strip-wound tape cores

For the last two materials, there are several variations of material available but the core shape (a toroidal tape-wound core) is set by the physical attributes of the material (brittleness). These properties also may limit the size of the core particularly with respect to the inside diameter. The choice for the various components for each function will be discussed separately.

2.3-COMPONENTS FOR POWER TRANSFORMERS

In power transformer functions, a large induction swing may be traversed calling for a core with highest effective permeability derived from the best core material and shape. A high saturation material with low losses under the operating conditions is desirable. While a toroid gives the highest effective permeability for a specific material, it is costly to wind. An ungapped ferrite E-core or other power shape is preferable for noncritical applications. For lower frequency power applications (line or mains frequency) laminated strip of iron or SiFe is commonly used. For somewhat higher frequency power applications, tape cores of thin strip SiFe, high permeability NiFe (80 Permalloy) high saturation NiFe (50 Permalloy) or CoFe (Supermendur) can be used. For high frequency operation, the workhorse of the power electronics components is the ferrite core that is available in many materials, shapes and sizes. Recently, some inroads have been made with the new amorphous and nanocrystalline materials.

Power transformers have one requirement in common, that is that the material saturation be as high as possible consistent with other factors such as core loss. As we shall see later, at very low frequencies, the materials are saturation-driven as eddy current losses are moderately low. These materials such as iron, low-carbon steel and heavy gage silicon iron will not be discussed in this book since they do not enter into the field of power electronics. At high frequencies, the materials are core-loss-driven so that materials such as ferrites are generally used. For medium high frequencies, materials such as thin gage silicon iron, amorphous materials and the new nanocrystalline materials are available. We normally make this choice on the basis of frequency of operation.

Vendors usually provide guidelines as to what materials are suitable for the various frequency ranges. The core losses are often given as a function of frequency. Although vendors generally list power materials separately, the user often has a choice of several available materials varying according to losses, frequency and sometimes, cost. Since power ferrites operate at the highest possible induction, we find, as we would expect, that they have the highest saturation of the ferrites consistent with maintenance of acceptable losses at the operating frequency. For frequencies up to about 1 MHz., Mn-Zn ferrites are the most widely used materials. Above this frequency, NiZn ferrites may be chosen because of their higher resistivities. Since ferrite cores make up the largest proportion of high frequency power transformers, the considerations for their choice will be discussed first.

2.4- FERRITE POWER TRANSFORMERS

Having mentioned that ferrite cores are the major components for power electronics we will discuss them first. Switching power supplies and ferrite expansion went hand in hand. To explain why ferrites were made to order for these applications, we must understand the implications of going to higher frequency operation. Ferrites have low saturation compared with most common metallic magnetic materials (such as iron) and also have much lower permeabilities than materials such as 80% NiFe. The low saturation of ferrites comes about from the fact that the large oxygen ions in the spinel lattice contribute no moment and so dilute the magnetic metal ions. This situation is compared to a metal such as iron where there is no such dilution. In addition, because of the antiferromagnetic interaction, not all the magnetic ions contribute to the net moment in ferrites but only those with uncompensated spins.

The first use of ferrite material in a power application was to provide the time-dependent magnetic deflection of the electron beam in a television receiver. The two ferrite components used were the deflection yoke and the flyback transformer. This application remains the largest in tons of soft ferrite used.

Another early use of power ferrites was in matching line to load in ultrasonic generators and radio transmitters. Ferrites were not considered for line power inputs because at the lower frequencies (50-60 Hz.),they were economically unattractive (lower B_{sat} and higher cost than electrical steels). However, today's ferrites are employed as noise filters in power lines on the input to all types of electronic equipment. The potential for using ferrites at high frequencies was always there but the auxiliary circuit components (mostly semiconductors) were not yet developed. In addition, earlier there was no great market or stimulus for high frequency power supplies.

One envisaged use was in high frequency fluorescent lighting at about 3000 Hz. This idea was suggested in the early 1950's but the need for setting up line power at these frequencies was never fulfilled. (See Haver 1976)

In the 1970's, the rapid growth of ferrites for use at high power levels occurred shortly after the similar growth of power semiconductors that could switch at very high frequencies. This design specifically required moderate cost magnetic components with low losses at higher frequencies and elevated power levels. Thus the age of the switched mode power supply (SMPS) was born. Coincidentally, the rapid growth of computers and microprocessors has required small, efficient power supplies that could be constructed with power ferrite components. The computer and allied markets are certainly providing much of the present day impetus for today's power ferrite development. The matching of SMPS component needs with ferrite properties is explained in the following two sections.

2.4.1- Frequency-Voltage Considerations

A changing electric current in a winding provides a corresponding change in a magnetizing field which sets up a resultant varying magnetic induction in a magnetic material. This changing flux will induce a voltage in another (secondary) winding. The general case of a voltage produced by a changing (not necessarily alternating) magnetic flux is given by Faraday's equation namely;

$$E = -N\, d\phi/dt = -N\, d(BA)/dt \qquad [2.1]$$

For a sine wave, the induced voltage is given by:

$$E = 4.4BNAf\mathrm{x}10^{-8} \qquad [2.2]$$

where: ϕ = Magnetic flux, maxwells
E = induced voltage, volts
B = maximum induction, Gausses
N = number of turns in winding
A = cross section of magnetic material, cm^2
f = frequency in Hz.

For a square wave, the coefficient is 4.0 instead of 4.44.

If we, for the present, minimize the effects of complicating problems such as core losses and temperature rise (which we will discuss later), we can use this important induction equation to examine the use of the variables in the most preliminary design. To obtain a given voltage with the most efficient arrangement, the tradeoffs can be as follows;

1. Increasing B by using a material with high induction such as 50% Co-Fe. This material is used in aircraft and space application where space and weight are important. However, there is a material limitation on how high the B can go. Ferrites may have saturation of 4-5,000 gausses. The highest RT saturation of about 23,000 gausses is found in 50% Co-Fe, so that there is a possible 4:1 or 5:1 advantage here for metals. See Table 2.1.
2. N can be increased which leads to higher wire resistance losses. Also, there is a maximum number of turns that can be wound around a core with a window or bobbin area. Using small wire size allows more turns, but the increased resistance (due to the increased length to cross sectional area) limits the useable current through the wire.
3. A can be increased. In addition to higher core losses, the larger cross-section requires a longer length of wire per turn leading to higher copper losses and a larger and heavier device. The larger cross section in a poor thermal conductor such as a ferrite also creates the problem of how to remove the heat produced in a large core. If the heat isn't removed, the temperature rise lowers the saturation induction, B_s of the ferrite. Under these conditions, if the induction-swing, ΔB, is large enough, the core may actually saturate and the current in the winding can become very large possibly causing catastrophic failure. This can damage the core, the winding and other components.
4. f can be increased. Here the effect can be quite dramatic depending on the frequency dependence of core losses. For instance, in going from a 60-Hz power supply to 100KHz supply, the factor is 1666. This coupled

with a 4:1 reduction in going from high B metals to low B ferrite still leaves 400^{+}:1 advantage. This permits a great reduction in the size & weight of the transformer, which reduces wire and core losses. In a high frequency power supply, increasing the frequency can exacerbate the thermal runaway problem if the exponent of frequency dependence is higher than that for flux density. We will deal with this subject with later in this chapter.

Table 2.1
Saturations of Various Magnetic Materials

Material	Saturation (Gausses)
CoFe (49% Co, 49% Fe 2 V)	22,000
SiFe (3.25% Si)	18,000
NiFe (50% Ni, 50% Fe)	15,000
NiFe (79% Ni, 4% Mo, Balance Fe)	7,500
NiFe Powder (81% Ni, 2% Mo, Balance Fe)	8,000
Fe Powder	8,900
Ferrites	4,000-5,000
Amorphous Metal Alloy(Iron-Based)	15,000
Amorphous Metal Alloy (Co-based)	7,000
Nanocrystalline Materials (Iron-based)	12,000-16,000

2.4.2-Frequency-Loss Considerations

We have shown that by increasing the frequency of a transformer, we can produce the desired voltage requirement at a greatly increased efficiency. However, we have neglected one consideration, that is, the increased losses that occur when we increase the frequency of operation. The additional losses incurred in the frequency increase are mainly eddy current losses caused by the internal circular current loops that are formed under ac excitation. The eddy current losses of a material can be represented by the equation:

$$P_e = KB_m 2f^2d^2/\rho \qquad [2.3]$$

where: P_e = Eddy Current losses, watts
K = a constant depending on the shape of the component
B_m = max induction, Gausses
f = frequency, Hz
d = thickness-narrowest dimension perpendicular to flux, cm
ρ = resistivity, ohm-cm

Again there is a trade-off for lower P_e. B can be lowered which means larger A to get the same voltage. Frequency, f, can be lowered which again means larger components. The thickness, d, can be made smaller, such as in thin metallic tapes, wire or powder. There are physical limitations to this variable, and also the high cost of rolling metal to very thin gauges.

The other measure we can take is to increase the resistivity. (See Table 2.2.) A comparison will demonstrate the advantage of ferrites. The resistivity for metals such as Permalloy or Si-Fe is about 50 x 10^{-6} ohm-cm. The resistivity of even the lowest resistivity ferrite is about 100 ohm-cm. The difference then is about 2 million to 1. Since the effect of the frequency on the losses is a square dependence and that of resistance only a linear one, the net effect on frequency is about 1400 to 1. Thus, losses to the 60 Hz operation, for the same size core, extend to 84,000 Hz, close to the 100 KHz we postulated for the voltage calculation. Granted this calculation is simplified, having omitted wire losses and loss differences due to B variations, but the order of magnitude is probably reasonable. In actual cases, 60 Hz power supplies operate at efficiencies of about 50%, whereas the ferrite high frequency switching power supplies operate at 80-90%.

Table 2.2
Resistivities of Ferrites and Metallic Magnetic Materials

Material	Resistivitity, Ω - cm
Zn Ferrite	10^2
Cu Ferrite	10^5
Fe Ferrite	4×10^{-3}
Mn Ferrite	10^4
NiZn Ferrite	10^6
Mg Ferrite	10^7
Co Ferrite	10^7
MnZn Ferrite	10^2-10^3
Yttrium Iron Garnet	10^{10}-10^{12}
Iron	9.6×10^{-6}
Silicon Iron	50×10^{-6}
Nickel Iron	45×10^{-6}

We must include another consideration in the comparison. We have mentioned the poor thermal conductivity of ferrite and ceramics in general. Aside from the difficulty of firing very dense, large, ceramic parts without producing cracks, there is also the previously mentioned problem of heat transfer. Because of this limitation, ferrite switching power supplies have not been made larger than about 10 KW. This is in comparison to the over 100 KW supplies that are made of metallic materials. However, since the large markets in power supplies are for home computers or microprocessors, and since these are well within the operational size of ferrites, there is no real size problem here. A comparison of magnetic properties of ferrites with other magnetic materials is given in Table 2.3. The advantages and disadvantages of ferrites as power transformer materials are given in Table 2.4.

Table 2.3- From MMPA(1996)

MAGNETIC MATERIALS						
		FERRITES	ALLOYS	IRON POWDER	AMORPHOUS
Initial permeability range		5 to 15,000	5000 to 300,000	5 to 150	3000 to 150,000
Curie temperature range	(°C)	100 to 500	500	750	210 to 485
Saturation flux density range	(k gauss)	3 to 5	8 to 24	10 to 12	5.5 to 10
Coercive force	(oersted)	.05	.003 to 3		.003 to 3
Loss factor $\tan \delta/\mu_i \times 10^6$					
	at f = 10 kHz	5	8	25	
	100 kHz	10	80	30	
	1 MHz	25	4000	100	
Resistivity	(Ω - cm)	10^1 to 10^8	10^{-5}	10^4	1.4×10^{-4}

Ferrites
(MnZn and NiZn)

ADVANTAGES:
- Highest frequency range.
- Least expensive.
- Bobbins & hardware readily available.
- Allow operation above audible frequency.
- Round winding areas.
- Easliy adapted to mass manufacturing.

DISADVANTAGES:
- Lowest flux density.
- Largest core needed below 20 kHz.
- Poor heat transfer.
- Low temperature range.

Table 2.4- Advantages and disadvantages of ferrites for SMPS transformer applications. From Bosley(1994)

2.4.3-Choosing the Best Ferrite Power Transformer Material

A material slated for a power application must meet certain special requirements. Although ferrites in general have low saturations, we must, at least, provide the highest available variety consistent with loss considerations. This is mostly a matter of chemistry. Along with this consideration is the need for a high Curie point. This generally means maintaining a high saturation at some temperature above ambient which approaches actual operational temperature. In addition to the saturation requirement, the material must possess low core losses at the operating frequency and temperature. The transformer losses, which include both the core loss and the winding loss, will heat the ferrite causing a reduction in the saturation to a value lower than that at room temperature. If this fact is not taken into account, the core may saturate at the higher temperature with disastrous results. A runaway heating situation could develop leading to catastrophic failure. Many ferrite suppliers have redesigned their materials such that the core losses will actually minimize at higher operating temperatures preventing further heating of the cores. The negative temperature coefficient of core loss at temperatures approaching the operating temperature helps compensate for the positive temperature coefficient of the winding losses in the same region. Roess(1982) has shown that the minimum in the core loss versus temperature occurs at about the position of the secondary permeability maximum. Thus, if the chemistry of the ferrite can be designed to have the secondary permeability maximum at the temperature of device operation of the transformer as described above, the core losses will also be lowest at that temperature. However, we must consider that this is only a local minimum. Having the minimum at 75-100°C.is a tremendous aid to the designer in avoiding thermal runaway, but still requires careful design work as the core loss increases above this minimum and the capacity for thermal runaway is still very real. Some smaller portable devices such as labtop computers operate intermittently at low ambient temperatures so materials with core loss minimum temperatures near room temperature may be used.

2.4.4-Power Ferrite Core Shapes

Power ferrites come in a variety of shapes. Although pot-cores were the ferrite shapes of choice in telecommunication ferrites, several required or preferred features for this application are not as critical in power usage. These include:

1. Shielding
2. Adjustability

In addition, pot cores are more costly and power ferrites must compete with other alternative materials. Therefore, shapes such as E cores, U cores, and PQ cores are more applicable to power application. Other shapes including solid center post pot cores can be used. The following chapter describes the types of shapes available. The shape of the core has a bearing on the amplitude permeability since the inductance is given by;

$$L = .4\ \pi\mu N^2\ l/A \qquad [2.4]$$

where: l = length of the winding, cm.
A = Cross sectional area, cm^2
L = Amplitude Permeability

Therefore, the longer the section on which the winding is placed and the shorter the height of the winding, the higher the inductance.

2.4.5-Component Processing after Assembly

After the ferrite component is wound, there are additional process steps that may affect the choice of component. These include;

1. Encapsulation
2. Soldering
3. Clamping
4. Winding
5. Gluing
6. Coating

Encapsulation, coating, gluing and clamping all put stresses on the core that may affect the magnetic properties. Some materials are more sensitive to stresses than others even though the properties are superior. Winding also stresses the core. Toroidal winding is costlier than bobbin winding. In modern printed circuit design, wave soldering is often used to attach components and leads to the board. Proper component choice will minimize the effect of the soldering temperature.

2.4.6-High Frequency Applications

Special attention must be paid if the frequencies of power supplies extend past 100 KHz and even to the 1 MHz region. First, the size of the core may be reduced significantly. Second, the core material must be modified to lower the core losses at these frequencies. The maximum flux density or B level used which, at lower frequencies, may have extended to 2000-2500 gausses may have to be reduced to something on the order of 500-600 gausses to attain the lower losses. The increase in frequency with smaller size and better efficiency may more than offset the lower saturation used. We will discuss designs at these higher frequencies at a later point in this chapter. Vendors design instructions are based on allowing the flux density to drop to lower values at higher frequencies in order to keep the core loss constant at 100 mW/cm^3.

2.5-METAL STRIP POWER TRANSFORMERS

In our discussion in Section 2.4.1, we found that a great advantage of some of the metal strip magnetic components was their high saturation, (Si-Fe, Ni-Fe, and Co-Fe). However, as we examined the core losses in Section 2.4.2, we also found that due to their low resistivity, the usefulness of these materials was diminished. Reducing the thickness or gage of these materials allowed them to operate at somewhat higher frequencies. For the 80% Ni-Fe, with a lower saturation, the very high permeability also reduced high-frequency losses especially in thinner gages. The Ni-Fe alloys had an additional advantage of being able to be reduced to an extremely thin gage (0.0005") Provided that the frequency is not too high, this material can be used in the form of tape cores, laminations an also cut cores. The advantages and disadvantages of Si-Fe and Co-Fe are given in Table 2.5 . The comparable listings for Ni-Fe are given in Table 2.6 . One general advantage of the metallic strip cores is that they can handle higher power levels than the ferrite cores. Aside from their higher saturations, their higher thermal conductivities allow them to dissipate heat more efficiently.

SiFe and Permendur
(3% SiFe and 50% CoFe)

ADVANTAGES:
- Highest flux density at low frequencies.
- Smallest size at a given frequency.
- High power handling capability.
- Highest possible operating temperature at >200°C.

DISADVANTAGES:
- Laminations must be assembled.
- Highest assembly cost for cut cores and laminations.
- Limited to frequencies below 2000 Hz.
- Permendur is very expensive.

Table 2.5- Advantages and disadvantages of SiFe as a SMPS transformer material From Bosley (1994)

2.6.-AMORPHOUS METAL STRIP CORES

A newer line of metal strip materials other than those described in the previous section are the amorphous metal alloys. The iron-based alloys have the high saturation for use as power transformers. Their resistivities are higher

Nickel Alloys
(50% NiFe and 80% NiFe)

ADVANTAGES:
- High flux density at lower frequencies (<10 kHz).
- Small size for a given frequency.
- High possible operating temperature (up to 200°C).
- High power handling capability.
- Unlimited size range for tape wound core.

DISADVANTAGES:
- Expensive materials.
- Need toroidal winding equipment.
- Normally used below 20 kHz.

Table 2.6-Advantages and disadvantages of NiFe alloys for SMPS transformer applications. From Bosley(1994)

than the crystalline magnetic alloys and they can be annealed for either low-frequency or higher frequency operation. However, they are magnetostrictive and are only used at lower frequencies. The Co-based materials have almost zero magnetostriction which gives them a high permeability and lower losses. The advantages and disadvantages of the amorphous materials are given in Table 2.7. In Figure 3.34 are shown the real and imaginary values of the complex permeability. Roess gives one advantage of ferrites over the amorphous materials in that they can only be produced in toroidal tape-wound cores.

Amorphous Alloys
(Fe-Based and Co-Based)

ADVANTAGES:
- High flux density for Fe-based.
- Very low core losses for Co-based.
- Frequency range up to 100 kHz.
- Unlimited size available as tape cores.

DISADVANTAGES:
- Expensive materials.
- Low flux density for Co-based.
- Co-based limited to temperatures below 100°C.

Table 2.7- Advantages and disadvantages of amorphous metal alloys for SMPS transformer applications. From Bosley(1994)

2.7. -NANOCRYSTALLINE-BASED POWER TRANSFORMERS

The newest of the soft magnetic strip power materials are the nano-crystalline. These materials were developed as an extension of the amorphous metal materials and are made by a similar process except that the material is annealed to produce a very fine grain size on the order of 10 nanometers. The iron based material has high saturation and permeability. The latter is due to almost zero magnetostriction similar to the Co-based amorphous alloys

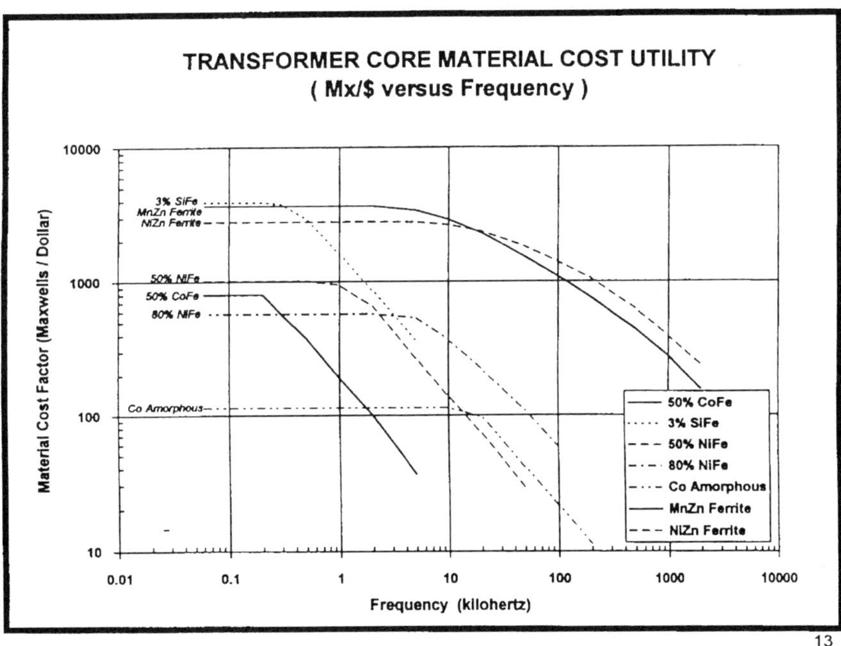

Figure 2.1- The maxwells of flux per dollar versus frequency for the various magnetic materials for SMPS transformer applications

2.8-COST CONSIDERATION FOR MAGNETIC COMPONENTS

Aside from the technical performance of the magnetic component, some consideration must be given to the comparative cost of the component. The final cost is determined by several factors;

1. Cost of raw materials
2. Cost of core fabrication
3. Cost of winding

Some of the advantages of a ferrite core are the low raw material and fabrication costs. Thin gage metals (SiFe, amorphous and nanocrystalline) have relatively high fabrication costs. For winding costs, toroids are the most expensive, then pot cores and finally, E-cores. Figure 2.1 gives the transformer cost utility in maxwells per dollar as a function of frequency. The predominance of ferrites especially at higher frequencies is evident.

2.9-COMPETITIVE HIGH FREQUENCY POWER MATERIALS

Roess(1987) has recently emphasized that a great virtue of ferrite power material is their adaptability, and even at higher power frequencies. He compares the losses of several competing power materials for the higher frequency operation. Trafoperm is a NiFe strip material. Vitrovac is an amorphous metal strip material and Siferrit is of course, a ferrite. The results are given in Figure 2.2 . Up to 100 KHz., the amorphous metal materials have lower losses than the ferrite especially the thinner gage type which remains lowest even at the higher frequencies. Roess points out that despite this disadvantage, a ferrite core is still the magnetic component of choice because of its much lower cost and its adaptability to be produced in many different shapes. The strip, on the other hand, has limitations on the shapes in which it can formed as shown in Figure 2.3. The new nanocrystalline materials were developed after this study.

A new fine-grained rapidly solidified nanocrystalline (not amorphous) strip material was introduced by Hitachi Ltd. It has much higher saturation (13,500 Gausses), higher permeability (16,000 at 100 KHz.) than ferrites and very low losses at 100KHz. Although these properties compare favorably to ferrites, it remains to be seen if the price and performance will allow it to compete with ferrites. In addition, the nanocrystalline cores have the same lack of fabrication versatility as other strip-wound cores

2.10-DESIGN CONSIDERATIONS IN COMPONENT CHOICE

Aside from the items previously mentioned in choosing a magnetic component for power electronics, there are additional considerations based on the other design requirements. These include;

1. Space, Volume and Weight Restrictions
2. Ambient Temperature-Heat removal
3. Environmental-Corrosion, Radiation, Vibration and Shock
4. Reliability-Lifetime
5. Regulation- Voltage variation
6. Safety considerations

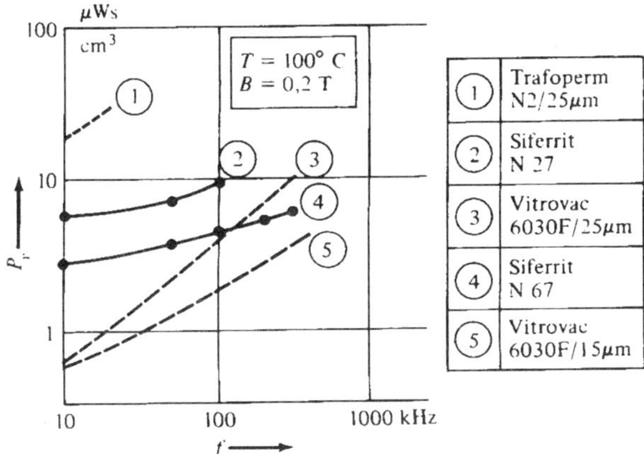

Figure 2.2-Core losses as a function of frequency for several different types of magnetic materials ;(1) is a NiFe metal strip material; (3) and (5) are amorphous metal strip materials; (2) and (4) are ferrites. From Roess(1987)

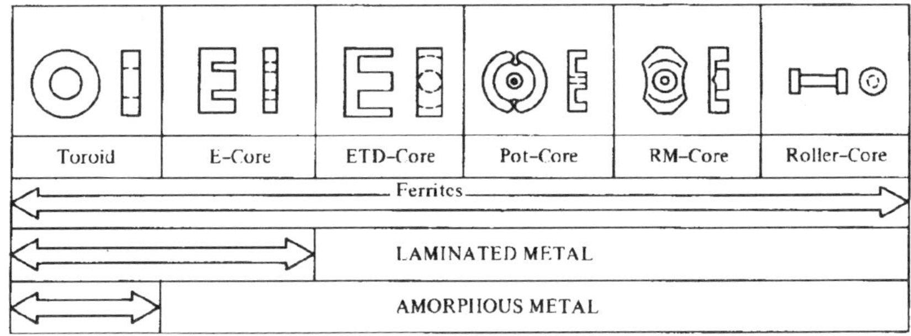

Figure 2.3-Formability of ferrites versus NiFe metal strip and amorphous metal strip. From Roess(1986)

2.11- FERRITES VS METALLIC MAGNETIC MATERIALS

There have been many comparisons of ferrites with other magnetic materials for power applications. The author (Goldman 1984) listed other metallic materials that were used for SMPS's. Goldman (1995) compared metal strip, powder cores and ferrites for various applications including power. Bosley (1994) presented a rather extensive study of the different materials for transformers and inductors versus frequency where the maximum flux was limited by saturation or core losses. The useable flux density under these limitations is given in Figure 2.4. For frequencies above 100 KHz., the MnZn and NiZn ferrites had the highest values. The performance factor in

Tesla-hertz vs frequency is shown in a figure in next chapter. Here again, above 100 KHz., the two ferrites were the highest. The economic trade-off for these materials is given in Figure 2.1 which charts the maxwells of flux per dollar as a function of frequency for the competing materials. Bosley (1994) also listed the advantages and disadvantages of the competing materials for SMPS transformers. Table 2.5 show these for SiFe and Permendur, Table 2.6 for the NiFe alloys, Table 2.7 for the amorphous alloys and Table 2.4 for ferrites. Snelling (`1996) presents a plot of the power loss density of power ferrite, Co-Fe amorphous metal strip and the Vitrovac 6030 nanocrystalline material vs frequency in figure in next chapter. The relative advantages in core loss depend on the frequency and flux density. The previous comparisons did not include the nanocrystalline materials that gained recognition shortly after the Bosley article even though Yoshizawa (1988) reported on them earlier. The frequency dependence of the permeability and loss factor of a nanocrystalline material was compared (Herzer 1997)along with those for a Co-based amorphous material and a MnZn ferrite. The permeability is higher and the loss factor is lowest for the nanocrystalline material. In addition the saturation induction was measured for the same materials (Herzer 1997). The nanocrystalline material has the advantage there. It remains to be seen whether the pricing of the nanocrystalline can be low enough to compete with the relatively inexpensive ferrite materials.

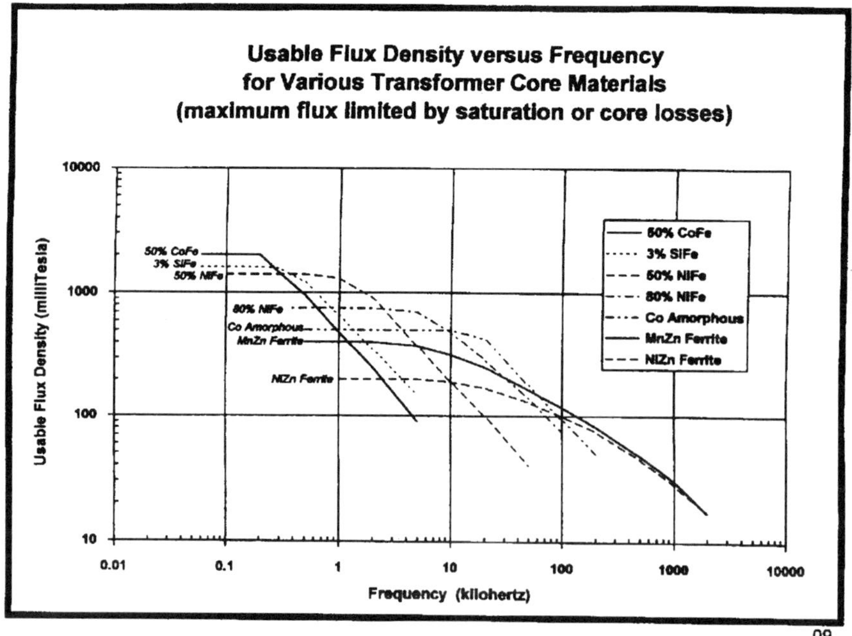

Figure 2.4-Usable flux density vs. frequency for various magnetic materials for SMPS transformer applications. From Bosley (1996)

2.12-OUTPUT POWER INDUCTORS

Power Inductors differ from the low-level inductors that we have dealt with in telecommunication applications. They are not used in LC circuits for frequency control. In power inductors, use is made of their ability to store large amounts of power in their magnetic field. As such, they can limit the amount of ac voltage and current. When this is done in the presence of a high D.C. current, the inductor, usually in combination with a capacitor, serves as a smoothing choke to remove the ac ripple in a D.C. supply. This is often done in the output circuit of the supply after rectification. Since there are large D.C. and smaller superimposed a.c. currents, they usually need gaps to prevent saturation. In addition to the increase in current and possible catastrophic failure at saturation, the incremental permeability drops close to zero and therefore, the required inductance specification is not met. With the gap, the magnetization curve is skewed to avoid saturation (See Figure 4.11). With regard to the ac component, the permeability of the gapped core is larger than one operating at saturation. The amount of gap depends on the maximum D.C. current, the shape and size of the core and the inductance needed for energy storage. The a.c. ripple is usually on the order of 10% of the D.C. signal. To estimate the maximum current, I_m, an extra 10% or more safety factor for transients is inserted in the design making I_m on the order of 1.2-1.3 I_o.

In some power inductor applications, as in the common mode choke, the magnetic core must sense the small difference between 2 magnetic currents and a high permeability toroid or ungapped shape must be employed. In some other of these inductor functions, the full power of the circuit passes through the magnetic component and some feature must be added to keep the core from saturating. The same is true for some energy-storage functions where a high DC current is present. In these two cases, either a core with a discreet gap (ferrite E core) or a distributed gap toroid (iron powder core) is warranted.

2.12.1-Ferrite vs Metallic Powder Inductors

Earlier in this chapter, we compared ferrite power transformer materials with their counterparts in metallic materials. For ferrite power inductors, the materials are mostly the same as the transformer materials. However in the case for metallic materials, the materials for power inductors are different than those used for power transformers. Whereas gapped ferrite cores are used for many power inductor applications, in the case of the metallic cores, the gap is a distributed one as found in powder cores. Bosley (1994) who did the analysis on SMPS transformer materials compared the materials for SMPS inductors in the same paper. The materials evaluated were;

1.NiFe Powder Cores- Molypermalloy and Hi_Flux Cores
2.Sendust Powder Cores- Kool-Mu and MSS cores
3.Amorphous Choke Cores
4.Powdered Iron Cores
5.Gapped Ferrite Cores
6, Metal Strip Cut Cores

Nickel Powder Cores
(MPP and High Flux)

ADVANTAGES:
- Low core loss at high frequency (MPP).
- Distributed gap, no fringing flux.
- Very high DC bias capability (High Flux).
- Available in wide variety of permeabilities, as high as 550µ.
- Very good temperature stability.
- Low magnetostriction.

DISADVANTAGES:
- Expensive.
- Only available as toroids.
- High Flux has moderate core losses.

Table 2.8-Advantages and Disadvantages of NiFe Powder Cores as Power Inductor Materials for SMPS Applications., From Bosley (1996)

Some of these materials were used for low level telecommunications applications. For power applications at high power levels, the materials may be somewhat different. Bosley (1994) listed the advantages and disadvantages of the above mentioned materials for power inductor applications. Table 2.8 lists these for NiFe powder cores, Table 2.9 for Sendust (FeAlSi) powder cores. Table 2.10 for amorphous metal choke cores, Table 2.11 for powdered iron cores, Table 2.12 for gapped ferrite cores and Table 2.13 for cut cores. Since power inductor often must operate under high D.C. bias conditions, the effective permeability for these materials are given. The DC bias curves for several of these materials are shown in Figure 2.5. Again, the nanocrystalline materials were not considered . Bosley also listed the core losses of the various inductor materials compared to ferrites in Table 2.14. Although the ferrites are lower in losses than the others listed. Bosley notes that, with a medium to large gap, there may be increased losses due to fringing flux This may increase ac copper losses near the gap. Nanocrystalline materials were not considered here as well.

Si-Al Powder Cores
(Sendust, Kool Mμ, MSS)

ADVANTAGES:	• Moderate core losses at high frequency. • Moderate price. • Distributed gap, no fringing flux. • Good temperature and frequency stability. • Very low magnetostriction.
DISADVANTAGES:	• Only available as toroids. • Winding costs higher than ferrites.

Table 2.9- Advantages and disadvantages of FeSiAl (Sendust) cores as power inductor materials for SMPS applications From Bosley (1996)

Amorphous Choke Cores

ADVANTAGES:	• Very high DC bias capability. • Flat permeability vs. DC bias curve. • Moderate core loss at high frequency. • Low magnetostriction.
DISADVANTAGES:	• Only available as toroids. • Higher price than ferrites. • Core must be boxed, can limit LI^2.

Table 2.10-Advantages and disadvantages of amorphous metal cores as power inductor materials for SMPS applications From Bosley (1996)

Powdered Iron Cores

ADVANTAGES:	• Most are low cost. • Distributed gap, no fringing flux. • Available in various shapes. • Available in large sizes. • Most economical for DC and low frequency applications.
DISADVANTAGES:	• High magnetostriction. • May need larger size when high frequency ripple is present.

Table 2.11-Advantages and disadvantages of iron powder cores as power inductor materials for SMPS applications From Bosley (1996)

Gapped Ferrite Cores

ADVANTAGES:
- Wide variety of shapes.
- Economical winding.
- Flat permeability vs. DC bias curve.
- Lowest core losses at high frequency.
- Widest choice of gapping.
- Low price.

DISADVANTAGES:
- Requires assembly.
- Lowest flux density.
- Large gaps may cause high copper loss due to AC fringing flux.

Table 2.12-Advantages and disadvantages of gapped ferrite cores as power inductor materials for SMPS applications From Bosley (1996)

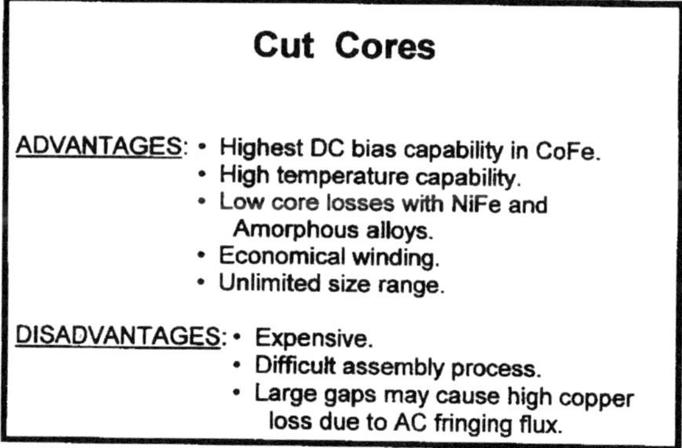

Cut Cores

ADVANTAGES:
- Highest DC bias capability in CoFe.
- High temperature capability.
- Low core losses with NiFe and Amorphous alloys.
- Economical winding.
- Unlimited size range.

DISADVANTAGES:
- Expensive.
- Difficult assembly process.
- Large gaps may cause high copper loss due to AC fringing flux.

Table 2.13-Advantages and Disadvantages of metallic strip cut-cores as power inductor materials for SMPS applications From Bosley (1996)

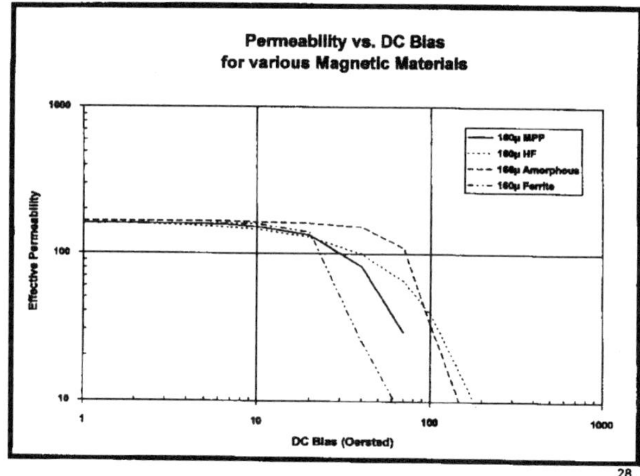

Figure 2.5-Effect of DC bias conditions on the effective permeability of several materials as power inductors for SMPS applications From Bosley (1996)

Inductor Core Loss Comparison

Ferrite used as base, 1×

Frequency	10 kHz	100 kHz	500 kHz	1 MHz
MPP	2×	5×	9×	12×
Sendust	2×	9×	18×	20×
Amorphous Choke	2×	15×	25×	25×
High Flux	10×	30×	80×	80×
Powdered Iron	20-40×	25-60×	25-100×	13-21×

NOTE: A medium to large size gap on a ferrite centerpost may cause total losses as high as the amorphous choke cores due to high AC copper losses caused by the fringing flux.

Table 2.14-Comparison of the core losses of various materials for power inductor applications versus gapped ferrites. From Bosley (1996)

Pauly (1996) reviewed the selection of a high-frequency core material for power line filters. He compared various powder core materials and gapped ferrites with respect to volume, sound level and cost. All cores were 1.84 inch toroids except the gapped ferrite which wa a gapped EC70/70G . The inductors were 4.0 mH. Ripple current was a 40 KHz. triangular Wave with peak-to-peak level of 33% of rated current. Output power was theoretical. Table 2.15 summarizes the results. The losses of the MPP, MSS (Sendust) and Hi-Flux cores were much lower than that of the powdered iron but the cost was dramatically lower. Best performance was found in the MPP cores. Table 2.16 is the author's opinion in the ranking of the cores as to the suitability of the various cores for a given application. In general smaller cores may be operated at higher frequencies and flux levels.

Core Type	Current Rating (A)	Output Voltage (V)	Output Power (kW)	Power Loss (W)	Power Loss (%)	Q@ 40kHz	ESR (Ω)	1K Price ($)
Mollypermalloy	6	320	1.92	11	0.6	74	14	10.97
MSS™	6	320	1.92	16	0.8	34	30	3.02
High Flux™	15	800	12.0	157	1.3	19	56	10.68
Powered Irom	13	693	9.0	131	1.5	16	64	.64
Gapped Ferrite	3.5	187	0.65	17	2.6	8	119	12.00
Ferrite/Litz	3.5	187	0.65	8	1.2	100	10	13.00

Table 2.15- Properties of Various Powder Core and Gapped Ferrite Materials as Output Choke Cores in Switching Power Supply Applications From Pauly (1996)

Core Type	Flyback Circuits	Line Filter	High Energy Circuit	Saturation Frequency Limit (kHz)	Filter Chokes	Precision Filters	Price
Molypermalloy	Good	Poor	Poor	3.5	Best	Good	High
MSS™	Fair	Good	Fair	1.6	Good	Fair	Low
High-Flux™	Poor	Best	Best	1.0	Fair	Poor	Medium
Powdered Irom	Worst	Fair	Good	0.8	Worst	Worst	Lowest
Ferrite/Litz	Best	Worst	Worst	20.0	Poor	Best	Highest

Table 2.16- Ranking of Cores in Author's Opinion for Suitability of Various Materials as Output Choke Materials for Switching Power Supply Applications- From Pauly (1996)

2.13- POWER FACTOR CORRECTION CORES

In some new ac input lines for switching power supplies, a "front end" boost pre-regulator is used to obtain an essentially Unity Power Factor or UPF. The circuit for this function is shown in Figure 2.6. An external logic circuit serves to control the duty cycle of the main switch, Q1 to raise the input voltage, V_i to the output voltage, V_o. Higher peak ac flux densities are present than in conventional output chokes so core losses are quite important here. If the wrong core or material is used, core losses will increase and the possibility of thermal failure may occur. As in the case of the output choke, the materials used are the various types of powder cores (iron, molyperm, high flux NiFe, Sendust) or a gapped ferrite. However, special considerations of the losses must be taken into account with iron powder cores so that larger cores and lower loss materials must be used.

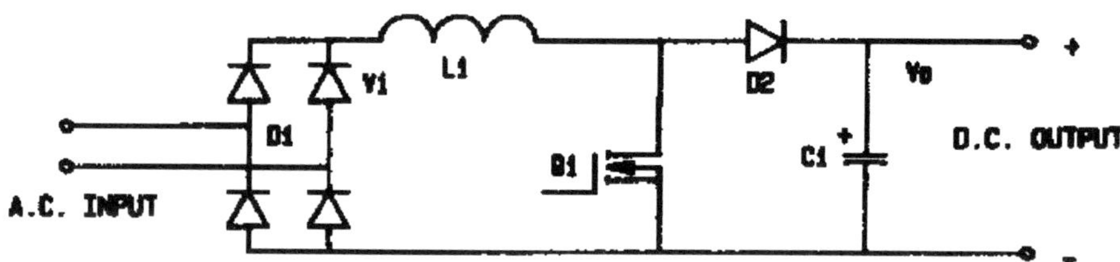

Fgure 2.6- A front-end boost preregulator for Power Factor Correction (PFC), (From B. Carsten, Application Note, Micrometals (2001)

2.14 -MAGNETIC AMPLIFIER CORES

In cases of multiple outputs in switching power supplies, their may be an imbalance in the output voltages. In cases where the regulation must be controlled very precisely, one solution is the use of a magnetic amplifier regulator circuit. The circuit for a forward converter with a 5V and 12V. ouput is given in Figure 2.7 . The 5V output uses PWM feedback circuitry to regulate. The 12 V output uses the magnetic amplifier or saturable-core reactor to regulate. The materials used for the mag amp must have high squareness or B_r/B_s ratios, low coercive force for small reset current and low core loss for small temperature rise. Components for this function are Square-loop NiFe tape cores, Co-based amorphous metal cores and square loop ferrite toroids.

Square Permalloy	
ADVANTAGES:	• Moderate price. • Highest flux density. • Low core loss below 100 kHz. • Low saturated inductance
DISADVANTAGES:	• Stress sensitive, must use a core box. • High core losses above 100 kHz.

Table 2.17-Advantages and disadvantages of square Permalloy as a magnetic amplifier for SMPS applications From Bosley (1996)

We have considered the use of square loop ferrites previously in conjunction with the ferroresonant transformer design. The advantages and disadvantages of the square permalloy are shown in Table 1.17.Those of the cobalt-based amorphous metal material are given in Table 1.18 and the corresponding ones for the square-loop ferrite are given in Table 1.19.

Cobalt-based Amorphous	
ADVANTAGES:	• Lowest core losses. • Moderate flux density. • Highest squareness. • Lowest saturated inductance. • Can be used encapsulated, without a core box.
DISADVANTAGES:	• Highest price. • Encapsulated version compromises the high squareness and the low saturated inductance. • Must be used below 100°C.

Table 2.18-Advantages and disadvantages of Co-based amorphous material as a magnetic amplifier for SMPS applications From Bosley (1996)

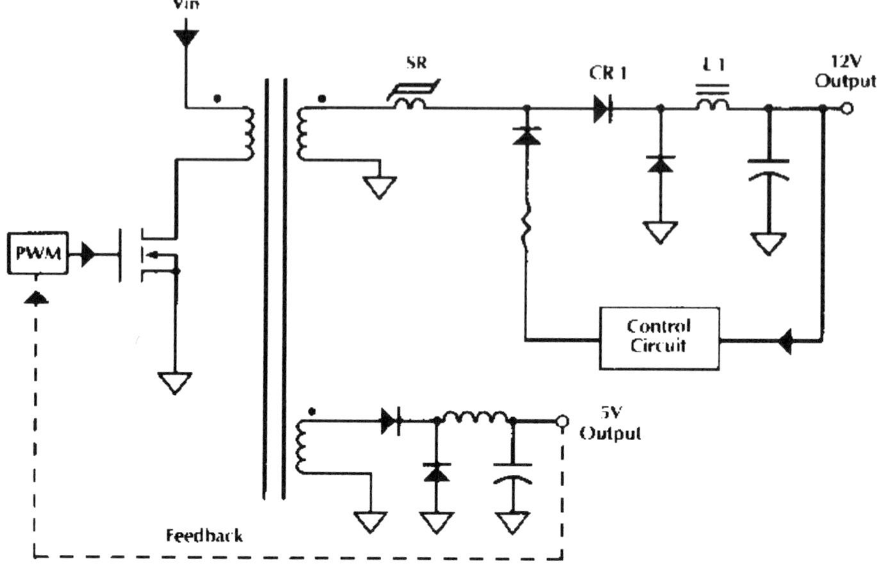

Figure 2.7 - Dual-output Magnetic Amplifier circuit.

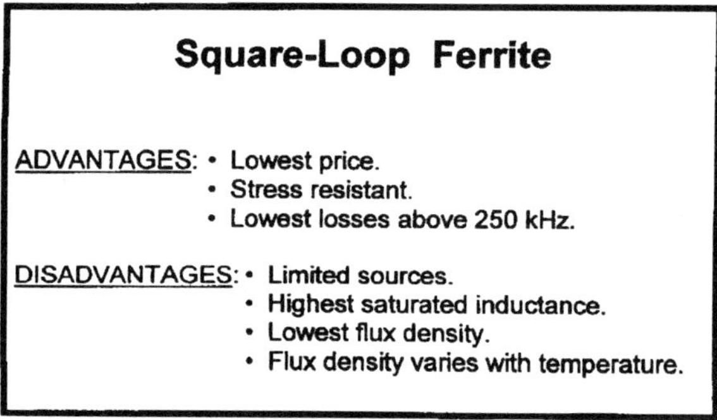

Square-Loop Ferrite

ADVANTAGES:
- Lowest price.
- Stress resistant.
- Lowest losses above 250 kHz.

DISADVANTAGES:
- Limited sources.
- Highest saturated inductance.
- Lowest flux density.
- Flux density varies with temperature.

Table 2.19-Advantages and disadvantages of square loop ferrite material as a magnetic amplifier for SMPS applications From Bosley (1996)

2.15-PULSE TRANSFORMERS

There are instances when the transistor switch supplying the square wave is triggered or fired by an external source or pulse generator. Ferrite cores, especially small toroids, are widely used in pulse transformers. This application requires transmission of a square wave with little distortion. The shape of a typical square wave voltage pulse is shown in Figure 2.8. During the time that the voltage pulse is on, the current is ramping up as is the flux density. In the case of a square wave, the ΔB is given in terms of the applied voltage, E, and the pulse width, T. For a specific core area and number of turns, the equation is;

$$B = ET \times 10^{-8}/NA_e \quad [2.5]$$

From the value of ΔB , the corresponding value of H, the magnetizing field can be determined from the vendor's curves on the material properties. From this value of H, and the l_e of the trial core, the excitation current can be determined from;

$$H = .4\,\pi N I_p/l_e \qquad [2.6]$$

If E, T, N are given and a ΔB is assumed, the effective dimensions of the core can then be given as;

$$l_e/A_e = 0.4\,\pi N^2 I_p\,\Delta B \times 10^8/ET\Delta H \quad [2.7]$$

The cores corresponding to various values of l_e/A_e are listed by the vendor. The pulse permeability is given by;

$$\mu_p L_p = ET/I_p \qquad [2.8]$$

The pulse transformers used in digital data processing circuits will usually have ΔB's on the order of 100 Gausses and are always little toroids inserted in small TO-5 cans for use on PC boards. Higher power pulse transformers may use pot cores or E cores that may be gapped to prevent saturation. All of the frequency related problems encountered in wide-band transformers are present in the pulse transformer, but here, it is evidenced by pulse attenuation (droop). As in the previous case, the permeability should be as high as possible, but when high pulse repetition rates or fast rise times are used, permeability concerns may be compromised for lower losses. Permeabilities of about 5-7000 are frequently used for small ferrite toroids.

2.16-COMPONENTS FOR EMI SUPPRESSION

Before our discussion of the actual components used for EMI suppression, it is useful to look at the circuitry involved along with the currents both intentional and unintentional. The latter, of course are the EMI interference or noise currents. Basically, there are two types of EMI currents, namely the common-mode and the differential currents. These are contained in a pair of wires leading to and from the load. The first of these is the differential currents whose current flow is the same as in ordinary intended or designed circuitry as shown in Figure 2.9. The differential EMI currents then flow in the same direction as the intended currents. If a current probe is placed around the pair of conductors, no current flow will be detected for either the intentional or unintentional (EMI) currents. In the case of the common-mode EMI currents, they flow in the same direction in both conductors. Now, while the dif ferential intended currents will cancel, the non-intentional (EMI) currents will not and there will be a current indicated with a probe. Another way of describing the two types of current flow is shown in Figure 2.9 that shows the voltages producing the currents. In the differential case, the voltage is between the high voltage line and the neutral line while in the common-mode case, it is the voltage between both the high voltage and neutral lines to ground. The components for EMI suppression extend from very small beads to rather large cable clamp cores. Some and even slug-type cores toroids are used in the coil type of suppressor

2.16.1-Materials For EMI Suppression

Until recently, the materials available for EMI suppression applications essentially were of two types. The most widely were and still are soft ferrites and the other less widely used one would be powder cores. Recently, amorphous and nanocrystalline cores have been used for the same purpose. Although sometimes the principal operational frequency of the circuit may be quite low (line or mains frequency, 50-60 Hertz), it is not primarily that frequency which is designed for in EMI suppression. It is rather the interference or disturbance frequency that mostly determines the choice of material used although the effect of the lower frequency (DC) must be dealt with in the design of the EMI filter. This interference frequency frequency can be high frequency ac or square or other digital waveform in the high Kilohertz or Megahertz region. The secondary consideration would be the operational frequency in that the material must pass the lower frequency with sufficient inductance. This means that, at low frequencies, the material must behave as a fairly good inductor but at high frequencies, it must be quite lossy. The frequencies involved in this application preclude any of the conventional metallic strip materials. Other possible new materials which will be listed later are the High Flux NiFe powder cores and the Sendust (Fe-Al-Si) powder cores.

2.16.2-Amorphous-Nanocrystalline Materials- EMI Suppression

One of the earliest uses of the amorphous material was for a choke coil that Toshiba called the “Spike-killer”. Presumably, only the cores are sold. The Fe-based amorphous materials used are under license from Allied’s

Metglas® Division (now Honeywell) . Vacuumschmelze does market a Co-based amorphous materials for EMI suppression applications. It is designated Vitrovac 6025 and is essentially a zero magnetostriction material. As an outgrowth of the amorphous materials, the iron-based nanocrystalline materials are the newest ones available and they have been used for EMI suppression. Their high permeabilities and low magnetostrictions made it very useful as a

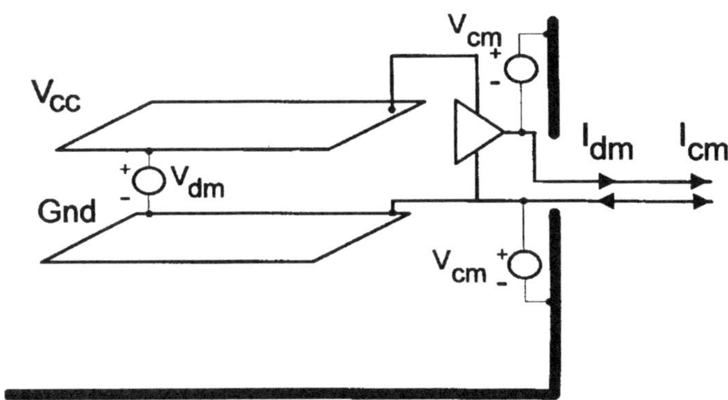

Figure 2.8- Common and differential mode currents in an EMI suppression circuit. From Lee Hill, (1994) MMPA Soft Ferrite Users Conference, Feb. 24-25, 1994, Rosemont, IL

common-mode choke at relatively low frequencies. The earliest nanocrystalline soft magnetic alloy was made by Hitachi and names "Finemet". The saturation is high but not in the Si-Fe class. The permeability is high but not in the 80 Permalloy class. The magnetostriction is low and about the same as the Co-based amorphous material. The resistivity is the same as the amorphous alloys but many orders of magnitude lower than ferrite. It is really the combination of most of the good attributes that make it attractive. High permeability materials without the high resistivities are useful for the inductive or permeability portion of the impedance and are limited to the lower frequencies

2.17-COMMON-MODE FILTERS

Since there are two types of EMI currents, there are two types of filters to handle them. The common mode filter uses a single core enclosing both conductors. The intended signal will be passed by cancellation of the two opposing currents. The EMI currents will produce magnetic flux in the core which because of the higher frequency will be attenuated. The common-mode filter is then transparent to differential mode currents and attenuates common-mode currents. One simple method of determining whether EMI noise is due to common mode or differential mode currents is by placing a ferrite core

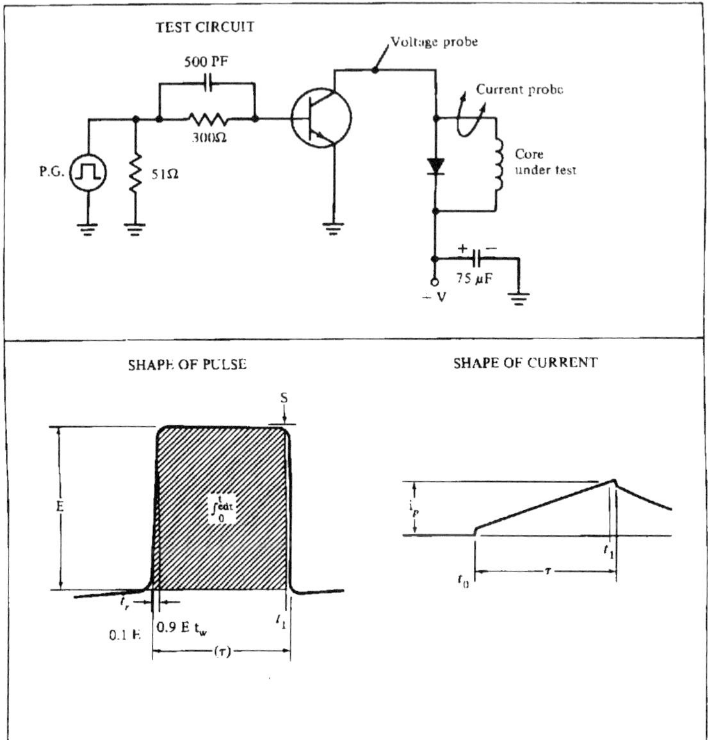

Figure 2.9- Shapes of voltage and current Wave forms in a ferrite core pulsed with a square wave (Courtesy of TDK Corp., Tokyo, Japan)

over both conductors, If the interference is removed or lessened, the noise was common mode. If no improvement, differential mode currents were the cause. Applications using the common-mode configuration are;

1. High current DC power filtering
2. Reducing noise on high speed differential lines

Very small common-mode currents(as low as 8μA) can cause failure in radiated or conducted EMI tests. For a differential mode filter, the equivalent current would be 19.9 mA, showing the greater sensitivity of the common-mode currents. Probably the simplest method of EMI suppression is the use of a ferrite core alone placed right over the device leads or on a PC board to prevent parasitic oscillations or attenuate unwanted signal pickup. The most common component of this type is the shield bead (sometimes with leads) but other shapes to accommodate cables (either flat) or round are available. For multiple lines, discs and plates are also used.

Ferrite beads are available to slide over conductors or they are available with preformed leads.

SUMMARY

This chapter has discussed the various magnetic components that are used for the different power electronic applications. The next chapter (Chapter 3) will concentrate on the available materials used for the appropriate components. The specific material properties of these materials will be discussed and shown how they are matched to the desired function and application. Chapter 4 will discuss the shapes of the appropriate materials. In Chapter 6 the required size will then determined completing the process of choosing the right component.

References

Goldman, A. (1985), Advances in Ceramics 15, Proc 4th ICF, p.421

Goldman, A. (1995), J. Mat. Eng. And Performance 4, 395

Haver, R.J.(1976), EDN,Nov.5,1976, 65

Magnetics (2000) Ferrite Core Catalog, FC601, Magnetic Div., Spang and Co, Butler, PA 16001

MMPA (!996) Soft Ferrites, A User's Guide SFG-96

Pauly, D.E (1996), PCIM (1996)

Roess, E.(1982), Transactions on Magnetics MAG18,#6,Nov.1982

Roess, E.(1986), Proc. 3rd Conf. on Phys Mag Mat. Sept.9-14,1986,Szczyrk-Bita, Poland, World Scientific

Roess, E.(1987) ERA Report-0285

Snelling E.(1988) Soft Ferrites, Properties and Applications Butterworths, London

Snelling, E(1989) presented at ICF5

Chapter 3
MAGNETIC MATERIALS FOR POWER ELECTRONICS

INTRODUCTION

In Chapter 2, the requirements of the various types of magnetic materials for power electronics were discussed. In this chapter, we will enumerate the specific materials under each category and list the properties associated with them. The categories are;

1. Soft ferrites
2. Powdered metal material
3. Metal strip

While a large percentage of materials for power electronics are now soft ferrites, there are new metallic material such as the amorphous and nanocrystalline type which may increase in importance and usage as further development and cost reduction occur.

Even within the group of soft ferrites, increasing frequencies may reduce the demand for the dominant manganese-zinc ferrites to nickel-zinc ferrites. In the past, the magnetic materials supplier community has been able to respond to changes in electronic design and system technology with new materials to meet the needs. There is no reason why it cannot do the same in the future.

The material properties that appear most often in magnetic component catalogs will be reviewed with respect to their significance and units. These properties are inherent or independent of their size or shape. In most cases, the measurements are made on toroids.

3.1-SOFT FERRITES FOR POWER ELECTRONICS

There are many properties that characterize a ferrite for power electronics applications. Some of them are;

1. Initial Permeability

2. Amplitude Permeability
3. Saturation Induction
4. Saturation vs Temperature
5. Core Loss vs Frequency
6. Core Loss vs Temperature
7. Core Loss vs Induction
8. Permeability vs Temperature
9. Permeability vs DC Bias
10. Curie Temperature
11. Resistivity
12. Coercive Force
13. Density
14. Complex Permeability
15. Remanence and Squareness
16. Performance factor, PF

These properties are ones that just define the material as opposed to ones that define the component. The latter properties will be discussed in a later chapter. Vendors of ferrite cores will mostly always show many of these parameters in their listing of material properties in their catalogs. These listings for the major suppliers will be shown in Chapter 6. An overall listing of magnetic properties of several classes of soft ferrites is shown in Table 3.1 .

3.1.1-Ferrite Material Permeabilities

Basically, the permeability, μ, is defined as the ratio of the induction, B, to the applied field, H. There are many different types of permeabilities as the above list shows. Some of the pertinent ones are initial permeability, amplitude permeability, maximum permeability, incremental permeability and complex permeability. Figure 3.1 displays a hysteresis loop describing several of the permeabilities. Most of these permeabilities can be derives by observing the appropriate slopes on the hysteresis loop.

Using the c.g.s. units of Gausses and Oersteds for B and H, the permeability, μ , is merely

$$\mu = B/H \qquad [3.1]$$

With MKS units in which B and H are given in Teslas and A/m, the permeability is given by;

$$B = \mu_r \mu_o H \qquad [3.2]$$

Where $= \text{where } \mu_o = 4\pi \times 10^{-7}$ henries/m

μ_o= permeability of vacuum

μ_r = relative permeability

Table 3.1
From MMPA, Guide to Soft Ferrites

TYPICAL FERRITE MATERIALS							
PARAMETER	MANGANESE-ZINC				NICKEL-ZINC		
Initial permeability range	*1600 to 3500	750 to 3500	3500 to 6500	6500 to 15,000	10 to 75	75 to 200	200 to 1000
Curie temperature range (°C)	180 to 280	150 to 250	130 to 150	120	250 to 500	200 to 500	90 to 200
Flux density range (gauss)	3500 to 5200	2600 to 4750	3000 to 5000	3000 to 5000	1000 to 2500	2500 to 3500	2800 to 3500
at H (oersted)	12.5	12.5	12.5	12.5	100	50	25
Power loss density (mW/cm^3) at 100 kHz - 1000 G	150						
Loss factor tan δ/μ_i x 10^6							
at f = 30 kHz				30			
100 kHz	10	5	20				
300 kHz							10-50
1 MHz		25				20-40	20-70
3 MHz						25-60	50-200
10 MHz					50-200	100	
30 MHz					100-1500		
100 MHz					400-2000		
**Typical frequency range kHz	10 to 500	10 to 2000	10 to 200	10 to 100			
MHz					10-150	2-30	1.5-5

Thus, μ_r in the MKS system is the same as μ in cgs. And μ_o is merely the conversion factor for B/H from MKS units to c.g.s. units. For a complete explanation of the various permeabilities, the reader is referred to the author's book, Handbook of Modern Ferromagnetic Materials, (Goldman, 1999).

3.1.1.1- Initial Permeabilities

The initial permeability is the limiting value of B/H as H approaches zero. For power applications, especially for output transformers or output chokes, the initial permeabilities of ferrites are in the region of 2000-2500. Although this particular permeability should apply only to low level operation and not for power, the temperature dependence of core losses follows inversely to the initial permeability (See next section). For common mode chokes and pulse transformers, the permeability of the ferrite used is high (5,000-10,000). For EMI applications, the permeability of the ferrites is

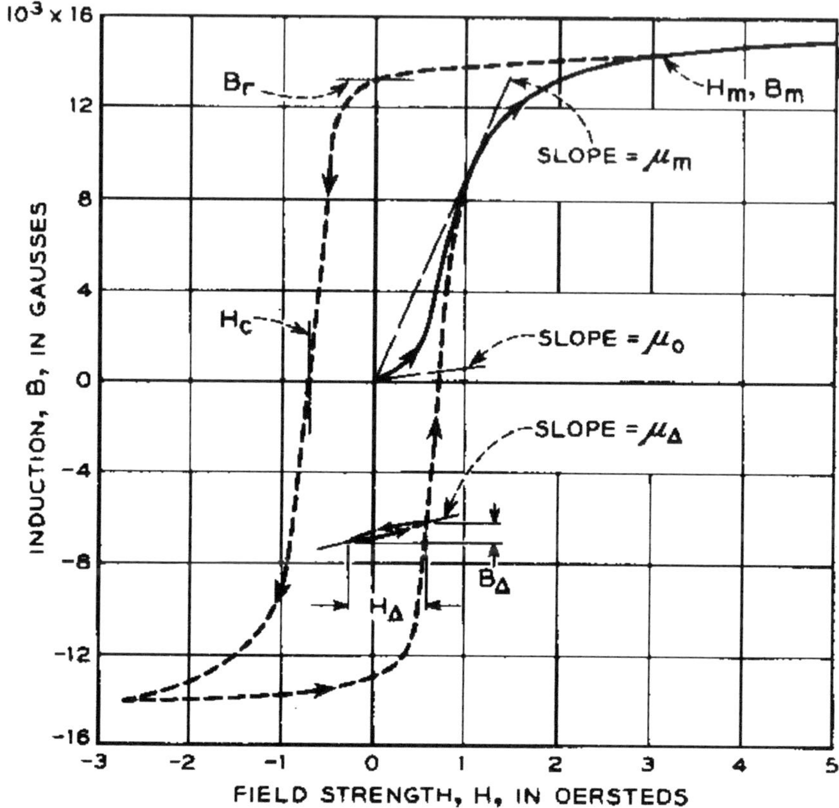

Figure 3.1- A Hysteresis Loop describing several different types of permeabilities, From Bozorth (1951)

mostly inversely proportional to the frequency of the disturbance. When this frequency is in the high Megahertz, cores with permeabilities as low as 14 are used.

3.1.1.2-Amplitude Permeability

The amplitude permeability is the relationship between flux density and higher field strength without the presence of a bias field. It is sometimes called the normal permeability. The permeability most applicable to power ferrites is the amplitude permeability which is usually shown in a plot of permeability vs ac induction amplitude. An amplitude permeability plot is seen in Figure 3.2. If the amplitude permeability is very low, then the magnetizing current or the number of turns must go up to provide the required H. This situation greatly increases the copper losses.

3.1.1.3- Maximum Permeability

The maximum permeability can be determined by noting the peak permeability on a plot of permeability vs flux density. The induction or flux density at the maximum permeability can be read from the graph. It is also the point of inflection on the magnetization curve for a material as shown in Figure 3.1 It is advisable to arrange for a ferrite output transformer to operate as close as possible to the flux density corresponding to maximum permeability subject to core losses.

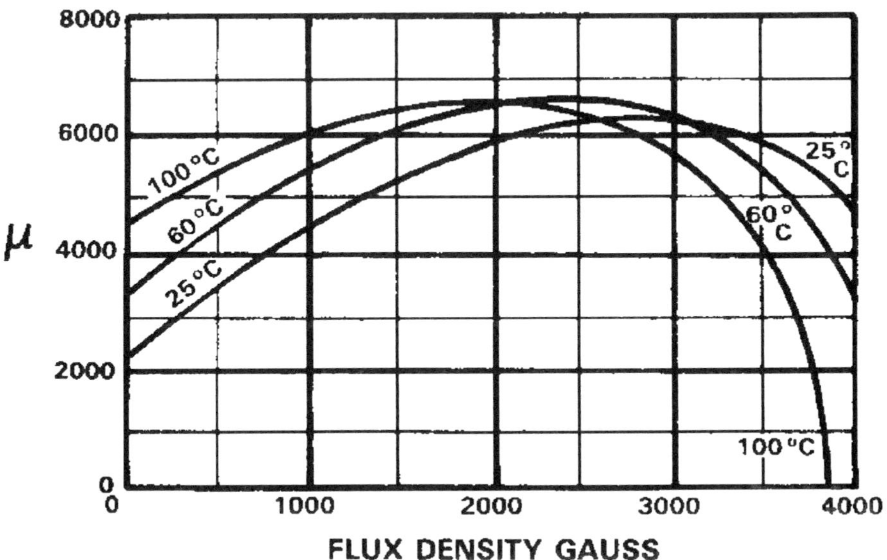

Figure 3.2- Amplitude permeability versus flux density for a typical ferrite power material at 25°C.,60°C. and 100°C.From Magnetics Catalog FC601(2000)

3.1.1.4-Permeability-Temperature Dependence

In MnZn ferrites, the permeability vs temperature curve usually shows a secondary permeability maximum (the first is just prior to the Curie temperature). See Figure 3.3. NiZn ferrites do not display the same effect.

As mentioned earlier, the initial permeability is not a directly governing factor in power ferrite choice. However, it has been shown that the temperature where the secondary permeability maximum occurs is generally

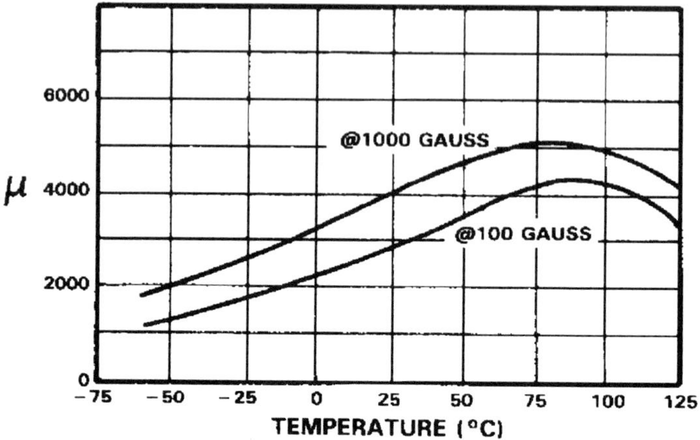

Figure 3.3- Variation of Permeability with Temperature for a typical MnZn Power Ferrite showing the position of the Secondary Permeability Maximum (SPM)

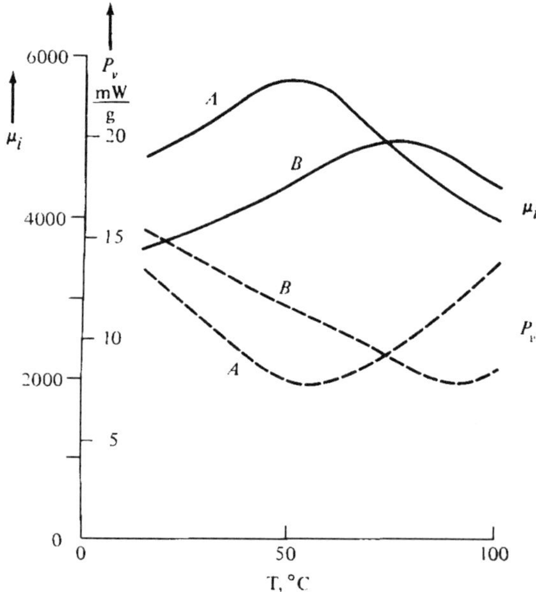

Figure 3.4- Temperature dependence of core losses for two different ferrite materials. The secondary maximum of the permeability and the minimum of the core losses occur at about the same temperatures for the respective materials. From Roess, Trans Mag. MAG 18 #6,Nov. 1982

in the vicinity of the minimum in core losses at a particular frequency. See Figure 3.4. In Figure 3.2 is shown the variation of permeability with B level. Thus, when the operating induction is very high, the permeability drops rapidly as we approach saturation, especially at 100°C.

3.1.1.5-Complex Permeability

The complex permeability which breaks down the permeability into the real and imaginary (loss) parts is usually applicable to low level or initial permeability operation. However, the use of Snoek's Law limiting the real permeability-frequency product is still applicable. It also shows the change from inductive to resistive impedence effects at the FMR (Ferromagnetic Resonance) point in EMI applications. A typical complex permeability graph is shown in Figure 3.5.

3.1.2-Saturation Induction(Flux Density) for Ferrite Materials

A material slated for a power application must meet certain special requirements. Although ferrites in general have low saturations compared to most metallic materials, we must, at least, provide the highest available consistent with loss considerations. This is mostly a matter of chemistry. Most good commercial Mn-Zn power ferrites have saturations or flux densities in the region of 5000 Gausses (0.5 Teslas). Those for high permeability ferrites for common-mode chokes are lower(3000-4500) with the emphasis on other factors to increase permeability. Nickel-zinc ferrites have saturations from 1000-3000 Gausses (0.1-0.3 T).

3.1.2.1- Temperature Dependence of Saturation

The saturation flux density of all magnetic materials decreases with increasing temperature. Figure 3.6 shows the saturation or flux density of a typical power ferrite as a function of temperature. In operation, the transformer losses, which include both the core loss and the winding loss, will heat the ferrite causing a reduction in the saturation to a value lower than that at room temperature. If this fact is not taken into account, the core may saturate at the higher temperature with disastrous results. A runaway heating situation could develop leading to catastrophic failure. Many ferrite suppliers have redesigned their materials such that the core losses will actually minimize at higher operating temperatures preventing further heating of the cores. The

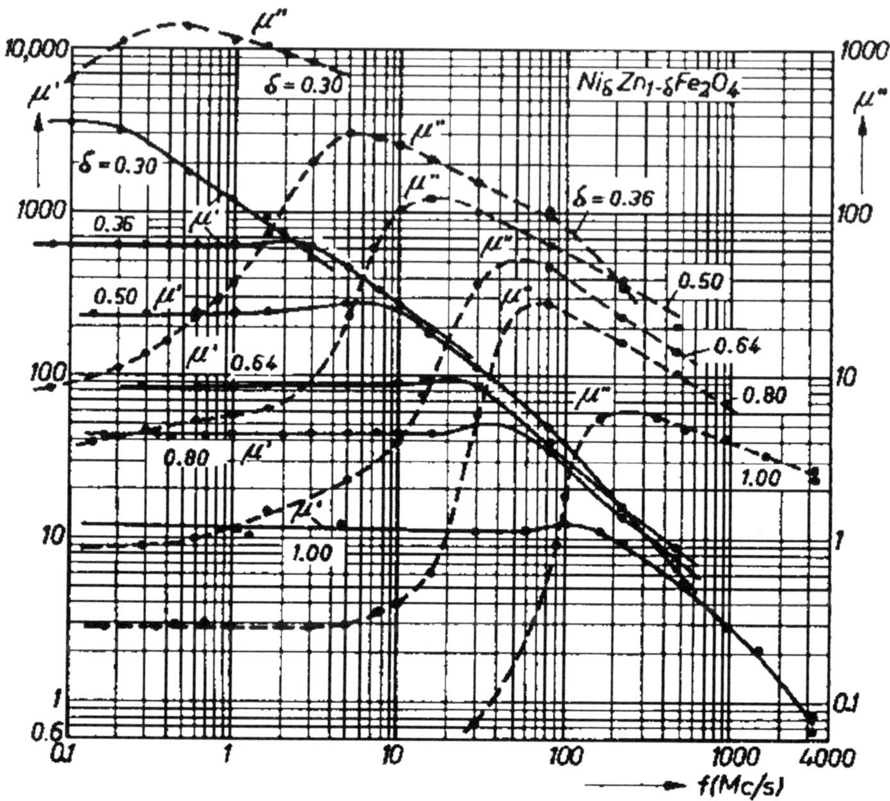

Figure 3.5- A Plot of Complex Permeability of Several Ferrite materials

negative temperature coefficient of core loss at temperatures approaching the operating temperature helps compensate for the positive temperature coefficient of the winding losses in the same region. Roess(1982) has shown that the minimum in the core loss versus temperature occurs at about the position of the secondary permeability maximum. Thus, if the chemistry of the ferrite can be designed to have the secondary maximum at the temperature of device operation of the transformer as described above, the core losses will also be low at that temperature. See Figure 3.4. However, we must consider that this is only a local minimum. Having the minimum at 75-100°C.is a tremendous aid to the designer in avoiding thermal runaway, but still requires careful design work as the core loss increases above this minimum and the capacity for thermal runaway is still very real.

Probably the most important feature common to power transformer requirements is the relatively high flux density. Power ferrites are usually

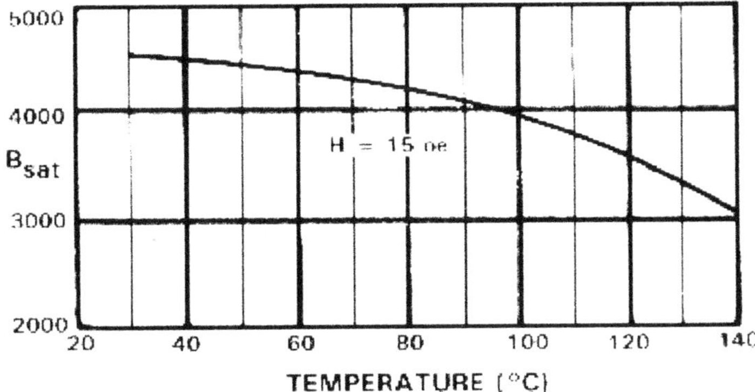

Figure 3.6 A plot of Flux Density (Gausses) versus Temperature for a Typical Power Ferrite Material

listed separately in vendors' catalogs and most show B_{sat} values of about 5,000 for materials to be used in the 25-100 KHz range. Earlier ferrite-cored power transformers operated at lower power levels and therefore, did not experience a large temperature rise. However, with present models, power levels are higher and temperature rises of about 40° to 60°C are not uncommon. The ambient temperature of the core without excitation is not considered room temperature (20°C.) but one much higher,(on the order of 40-60°C.) because of the heat generated by other components. The temperature rise of the transformer is added to the elevated ambient. At these higher operating temperatures (80°-120°C) the saturation of the material will drop as shown in Figure 3.4. However, when the saturation is high, the Curie point will usually be correspondingly high so there is little likelihood of catastrophic failure with moderate caution in design. The reduced saturation must be considered in the choice of B_{max} (operating) and is the one used in the induced voltage calculation. Figure 3.7 shows the reduction in theΔ B in a power transformer as the operating temperature is raised.

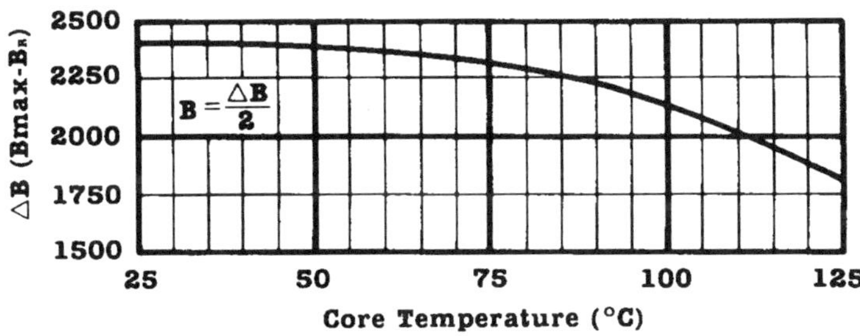

Figure 3.7-Decrease of ΔB with temperature to maintain lower losses in a unipolar driven transformer. From Martin (1987)

3.1.3-Core Losses

Another property of the material that must be examined is the core losses at the frequency of operation. Here again, vendors' curves will show losses as a function of frequency and flux density. In recent years, the switch to higher operating temperatures has prompted the development of materials whose losses are lower at the elevated temperatures of operation. The difference between the older material & the newer one is shown in Figure 3.8. If a

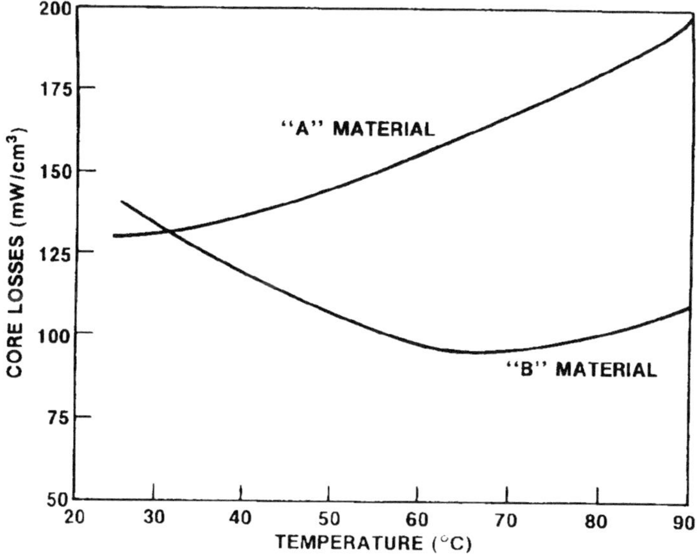

Figure 3.8-Temperature dependence of core losses for two different ferrite materials. A material has lower losses at room temperature but the B material losses are lower at 60°C., the temperature that the transformer might be expected to operate

higher frequency is chosen, the previously used flux density may have to be reduced to keep the losses from becoming excessive. The units for the core losses are given in units of mW/gm (W/kg) or mW/cm^3. If the units are given in terms of mW/cm^3/cycle, the value must be multiplied by the frequency. For metals, the units for these losses are watts per pound, (or W/Kg). For ferrites, the units usually used are mW/cm^3 and are often measured at a higher than room temperature (usually at the temperature of intended use). The reason for the difference in units between metals and ferrites lies in the fact that each product appears more favorable in the units chosen. The densities for the two materials make the difference. For example, let us assume that the core loss for the metal product is 1000 W/Kg. and that of the ferrite is 1200 W/Kg. Clearly, the metal is better. If we convert the units to mW/cc, the calculations are;

For the metal material with a density of 8 g/cc;

1000W/Kg = 1W/g x 8 g/cc = 8 W/cc = 8000 mW/cc

For the ferrite material with a density of 5 g/cc;

1200W/Kg = 1.2W/g x 5g/cc = 6 W/cc = 6000 mW/cc

In the changed units the ferrite appears superior.

A typical plot of losses versus f and B is shown in Figure 3.9.

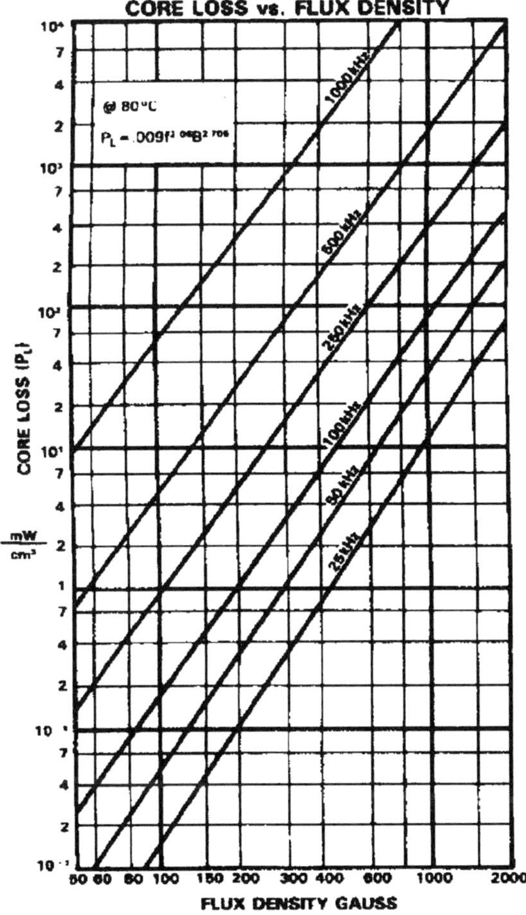

Figure 3.9- Core loss curves of a power ferrite material as a function of flux level and frequency.

McLyman(1982) has used an empirical equation similar to the Steinmetz equation to correlate losses with f and B. The applicable equation is

$$P_e = kf^mB^n \qquad [3.3]$$

where: P_e = core loss in watts/Kg
f = frequency, Hz
B = Flux density, Teslas (10^4 Gausses)

For the material shown in Figure 3.9, the equation would be;

$$P_L = .009f^{2.06}B^{2.705} \qquad [3.4]$$

For power ferrites, the value of m is on the order of 1.5 while that of n is about 2.5. At higher frequencies, n will increase possibly due to the onset of ferromagnetic resonance. The higher value of the B exponent, n, over the f exponent, m, indicates that the losses are more sensitive to variations in flux density than in frequency. This equation is valid over a rather limited frequency range. Strictly speaking, this equation should only represent the hysteresis losses with the eddy current losses given by Equation 2.2. However, since the eddy current losses are low in ferrites, this equation may be used. over the limited frequency range. We will discuss the importance of these coefficients of f and B later in this chapter. Snelling (1996) has shown the progression of ferrite materials with the year and frequency. See Figure 3.10 As the frequencies of operation for the SMPS have increased, new materials have been developed concurrently to operate under the new conditions.

3.1.4-Output Power Considerations-Performance Factor

Stijntjes(1989) has proposed a new material criterion for power transformer ferrites. He cites an expression by Mulder(1989) for the throughput power;

$$P_{th} = W_d \times C_d \times f \times B \qquad [3-5]$$

Where P_{th} = Throughput power, Watts
W_d = Winding design parameter
C_d = Core design parameter

Using a reasonable figure of 200 KW/m^3(mW/cm^3) proposed by Buthker (1986) for the allowable losses to maintain the required maximum temperature rise, Stijntjes (1989) develops a factor which he calls PF_{200}. This is the product of the material-dependent parameters in the equation (namely f and B) which will yield losses of 200 mW/cm^3. The other factors that determine

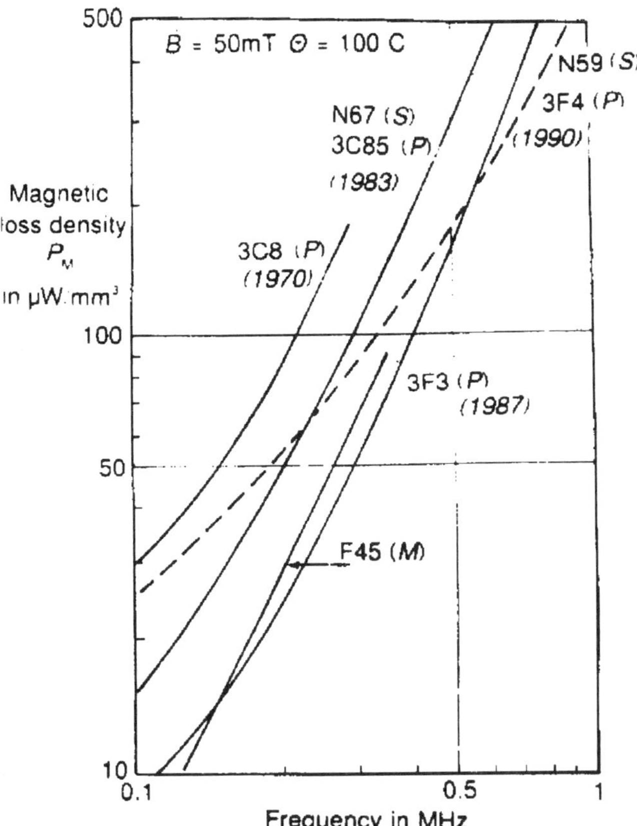

Figure 3.10-Magnetic loss density versus frequency for successive developments of MnZn power ferrites

the throughput power are related to winding and core design. Figure 3.11 shows a plot of the PF factor for losses of 100 mW/cm^3 for several transformer core materials.

3.1.5- Curie Temperature

For power ferrites, the need for a high Curie point goes along with the need for high saturation. Since the operating temperature is also high to obtain the maximum available power, a material with a high Curie point is generally chosen. Curie points for power ferrites are generally over 200° C. Curie points for high permeabilities are quite low with those for 10,000 perm materials in

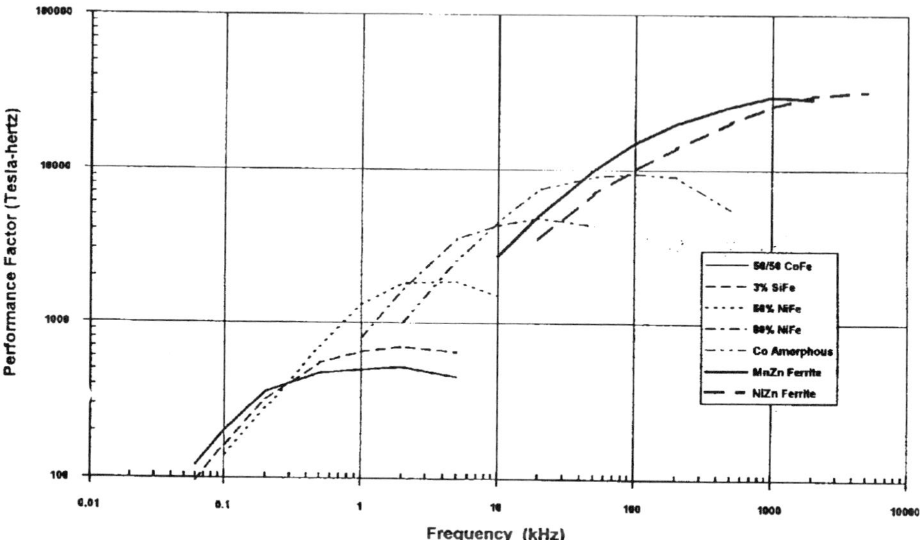

Figure 3.11- Performance factor in Tesla-hertz vs frequency for several magnetic materials for SMPS transformer applications From Bosley (1996)

the neighborhood of 110-120 ° C. Fortunately, for common-mode chokes, the power level is low and thus the cores do not get very hot.

3.1.6-Resistivity

One of the principal advantages of ferrites for operation at high frequencies is their high resistivity compared with metallic magnetic materials. However, even in ferrites, there is a wide spread of resistivities possible. The very lowest resistivity in ferrites is found in magnetite, a naturally occuring material that has little commercial importance. The high saturation ferrites with a large excess of iron or magnetite in its chemistry have the lowest of the commercial MnZn ferrites(100-200 Ω-cm). Power ferrites have somewhat higher resistivities (200-2000 Ω-cm.) with the highest values used for the highest frequency operation. The resistivity of a ferrite is dependent on the frequency and temperature. The resistivity decreases with both temperature and frequency. The resistivities for various magnetic materials are given in Table 2.2.

3.1.7- Density

Although metallic magnetic materials have mostly higher saturations than ferrites, the density of ferrites(4.8-5.0) are lower than metals(> 8.0), the weight of a ferrite is lighter for the same volume of component. MnZn ferrites have densities from 4.7-4.8 g/cm^3 while NiZn ferrite densities are in the range of 4.7-5.2 g/cm^3 .

3.1.8-Remanence and Squareness

The remanence or retentivity is the B value remaining when the material has been driven into saturation and the magnetizing field is reduced to zero. For conventional power materials, the remanence is fairly low. Squareness or B_r/B_s ratio does is not specified except for materials for magnetic amplifier or resonance converter applications. Most power ferrites have rounded hysteresis loops.

3.1.9 Coercive Force

Coercive force is the reverse field necessary to reduce the induction of a magnetic material from saturation to zero. For soft magnetic materials including power materials, the lower the coercive force, the lower the loss. When the frequency of operation of a material is increased, the coercive force will increase widening the loop and this increasing the loss.
design.

3.1.10- Ferrite Power Inductor Materials

Since there is always a large DC component in power inductors, there must be provision for preventing the core from going into saturation. In ferrites, this is accomplished by the use of an air gap. Since the ac component is small (on the order of 10% of the total), core losses are not as important as in the case of power transformers. The materials for ferrite power inductors have traditionally been the same as those used for transformer applications although some recent materials offered by Epcos claim to have superior power inductor performance than their transformer materials. Figure shows the DC bias of the new material, N92, compared to the older one, N87. The rated current of output chokes can be increased by 25% using this material. Figure

3.12 shows the variation of permeability as a function of DC Bias of a gapped ferrite vs other magnetic materials.

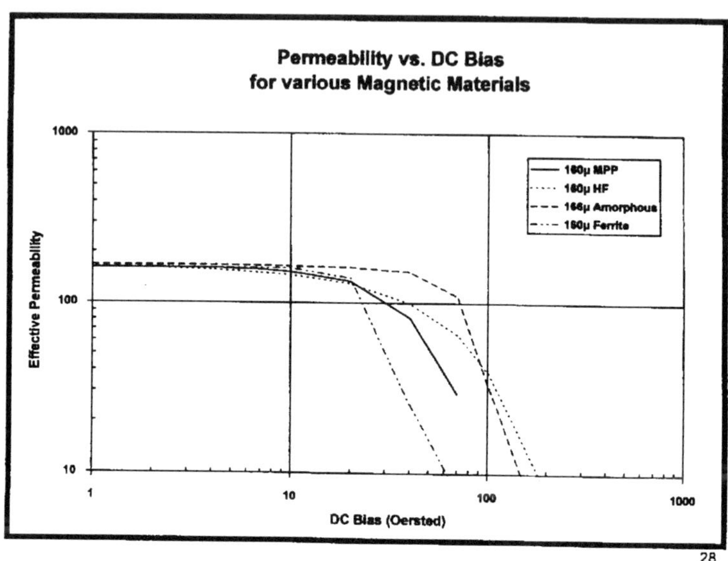

Figure 3.12-A graph of permeability vs D.C. bias for a gapped ferrite core and several other materials (Bosley (1994)

3.1.11-Magnetic Properties- Ferrite Common-Mode Materials

In the case of ferrite common-mode chokes, there is no DC present and the magnetic core and since the core encircles both conductors, the ac component is only the small difference current. Therefore, to achieve a satisfactory, inductance, a high permeability material is needed. Ferrite materials for common-mode chokes normally have permeabilities of from 5,000 to 15,000 depending on the frequency of the disturbance. The suppression is determined by the impedance of the circuit which can be inductive or resistive. Figure 3.13 shows the impedance of 3 common-mode choke materials. Their permeabilities for J, W, and H are 5,000, 10,000 and 15,000 respectively. At low frequencies, the impedance is inductive with little resistance. At higher frequencies, the resistance becomes more important as the impedance peaks. The higher the permeability, the lower the frequency at the peak. Figure 3.14 shows the resistance of the same 3 materials.

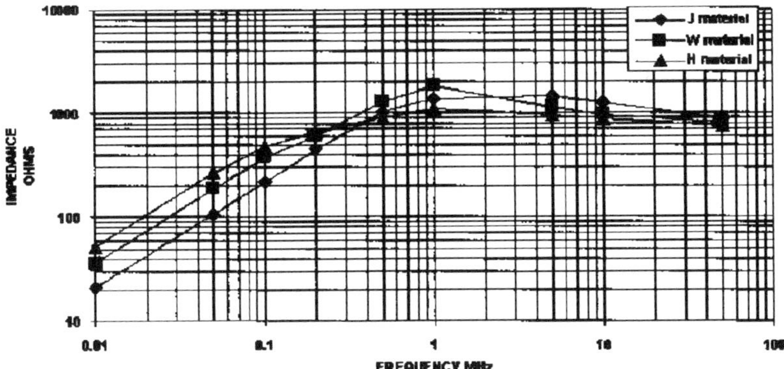

Figure 3.13 - Impedance vs Frequency for three common-mode choke materials

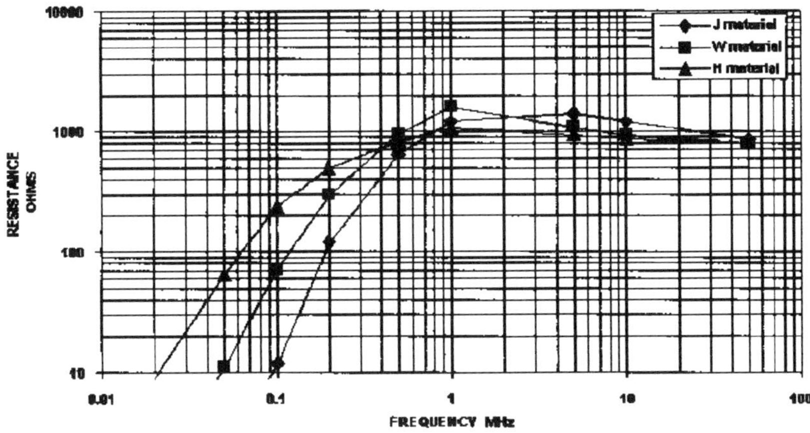

Figure 3.14 -Resistance vs Frequency for the 3 common-mode choke materials of Figure 3.13

For interferences at higher frequencies, materials of lower permeability may be required to shift the peak impedance to higher frequencies.

3.1.12-Magnetic Properties of Ferrite EMI Materials

The ferrites used in EMI applications are of 2 generic types. i.e., NiZn and the MnZn. When one looks at the technical papers on ferrite materials for applications such as low-level transformers and inductors or for high level transformers and inductors, there is a large number of these giving close details on the chemistry and microstructure. Papers on the chemistry of EMI suppression materials are notoriously absent. Part of the reason for this anomaly may be due to the recent recognition of their usefulness in combating EMI. However, more to the point, it maybe due to the fact that many companies have specific proprietary chemistries and processing to obtain the good EMI material. Another factor may be the small number of companies who do a sizeable business in this application. It is pretty clear that, in the U.S., two companies have dominated the field. They are Fair-Rite Products and Steward. This is somewhat surprising since a large market for these EMI components is in Europe or for products made elsewhere but sold in Europe.

Figure 3.15 shows the real and imaginary permeabilities of a EMI material. Figure 3.16 gives the impedance, inductance and resistance of an EMI material. Figure 3.17 shows the frequency depence of the impedance for a wide range of EMI materials.

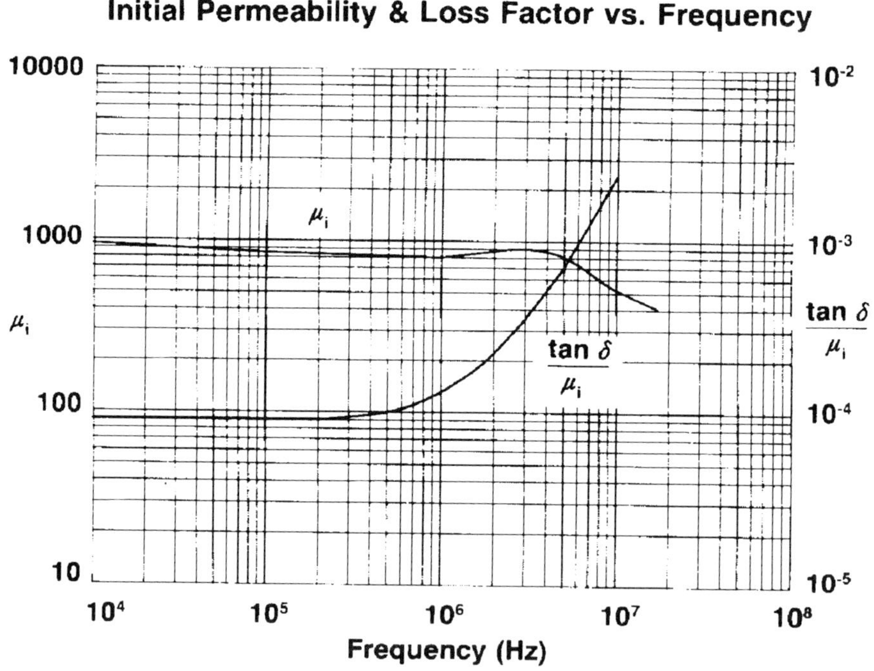

Figure 3.15- Frequency dependence of μ' and Loss Factor ($\tan\delta/\mu'$) for a EMI suppression ferrite.

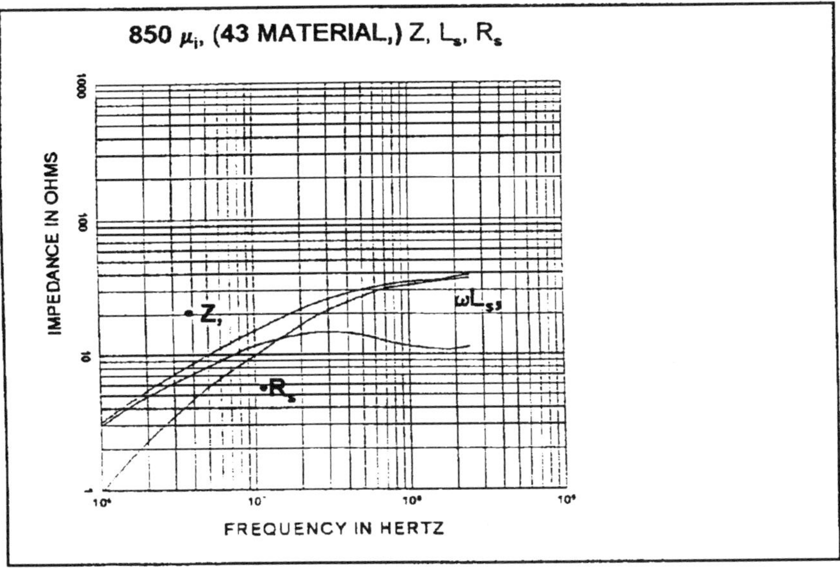

Figure 3.16-Frequency dependence of X_L, R, and Z for an EMI suppression material. From Parker (1994) Courtesy Fair-Rite Products

Since the frequencies for many EMI suppression applications are in the higher Megahertz band, the ferrites for EMI suppression are mainly of the NiZn variety. The use of MnZn material has been limited to the high KHz. or, at most, the very low MHz.region. The MnZn materials that are often used for noise filters at lower frequencies are the high permeability (5000-10,000 perm) ferrites. These materials have been described in an earlier chapter. A recent MnZn material typified by the Philips 3S4 material has been introduced to compete with the NiZn material in the 800-1200 perm range and at somewhat higher frequencies. The resistivity of this material is said to be 2 orders of magnitude than other MnZn materials.The NiZn materials normally contain copper for lower temperature firing and Mn for increased resistivity. The permeabilities for many of these materials are in the 500-1200 perm range. There is also a NiZn material with cobalt additions for higher frequency operation. This material is of lower perm (125). In addition to the chemistry of EMI suppression material, there is also concern for the microstructure.

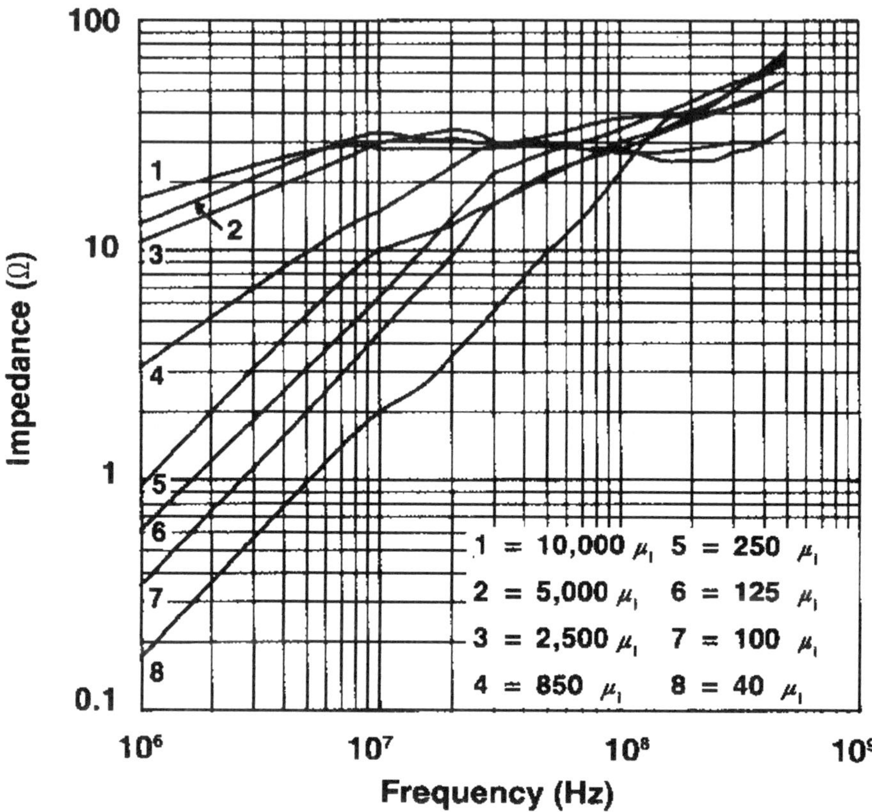

Figure 3.17- Impedance versus frequency curves for a wide range of EMI suppression materials. Courtesy Fair-Rite Products

When high perm MnZn material is used, large grains are conducive to high perm and high inductance needed to pass the lower frequencies but cause permeability fall off and rapidly increasing losses at somewhat higher frequencies where the suppression is desired. I would guess that the higher frequency MnZn material has increased resistivity due mainly to the prominent grain boundary structure. For the higher frequency, finer grain size and various degrees of intentional porosity aid in passing the medium high frequencies while attenuating the very high frequencies. Obviously, there is an interplay between chemistry and microstructure to obtain the desired result. Another consideration in these two property-determining factors is the influence of DC bias on the EMI suppression. Much of this information is obtained by experience and experiment.

3.1.13-Frequency Characteristics of EMI Materials

The function of the EMI suppressor core must meet two general needs, namely;

1. The passage of the lower frequency with sufficient inductance.
2. The suppression of the higher frequencies.

One governing rule of EMI suppression is Snoek's Law;

$$f = \gamma M_s / 3\pi(\mu - 1) \text{ Hz.} \quad [3.6]$$

Where f = Resonant frequency

M_s = Saturation Magnetization

γ = Gyromagnetic Ratio

This equation can also be approximated by

$$f = B_s/\mu_i \text{ MHz.} \quad [16.2]$$

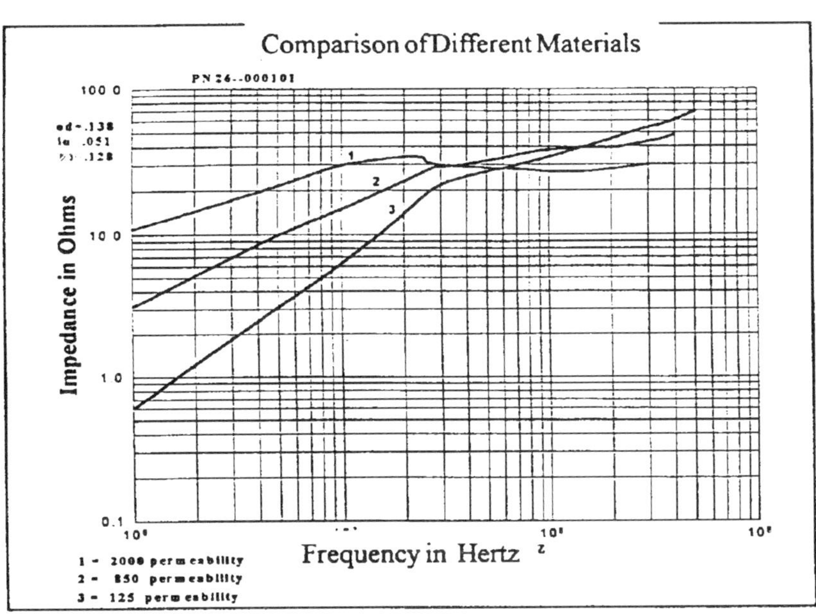

Figure 3.18-Impedance versus frequency curves for EMI materials in the 10-100 MHz. Range. From Parker (1994) Courtesy Fair-Rite Products

This shows what we have already surmised, that is, when f is low, the permeability will be high. When f is high, μ must be low. Thus, the higher the frequency of operation, be it ferrite or powder core, the lower the perm, and the converse. It also follows that the higher the perm, the lower the frequency of fall off and vice versa. Figure 3.18 shows the frequency characteristics of EMI materials in the 10-100 MHz. range.

3.2-MAGNETIC PROPERTIES OF METALLIC STRIP MATERIALS

Some of the magnetic properties of the metallic strip materials are similar to the ferrite materials. There are fewer chemistries for the metallic materials but, in contrast to the ferrites, some of their properties such as permeability and losses depend on the thickness or gage of the material. Therefore material specifications are specific for each gage. The only metallic strip materials that can compete as power electronic applications even at the lower frequencies are the very thin gages from about 0.005" to about 0.006 ". The core loss curves for Alloy 48 (a non-oriented 50% NiFe alloy) Orthonol (an oriented square-loop 50% NiFe alloy, Permalloy 80 an 80% NiFe alloy and Supermalloy, a very high permeability 80% NiFe alloy are given in Figures 3.19-3.22.

3.2.1-Permeability of Metallic Magnetic Strip Materials

In the case of magnetic metallic strip materials, the permeability most often specified in catalogs is the impedance permeability, which is equal to B/H where H is determined from the total magnetizing current and B is deternined from the voltage produced across the coil. It differs from other permeabilities in that it takes into consideration the magnetizing current which is a result of both the core losses and the inductive reactance (or material permeability of the inductor. It is measured by placing a voltmeter across the inductor and an ammeter in series with it. Using the standard B and H equations, the permeability is calculated.

The highest permeability and lowest losses at high frequency are found in the 80% NiFe alloy, Supermalloy. The permeability is either measured at D.C. or some higher frequency.

3.3-AMORPHOUS METAL MAGNETIC MATERIALS.

For many years, it had been predicted by many magneticians (including the author) that amorphous magnetic materials could only be paramagnetic. During the 1960's,small laboratory specimens of weakly

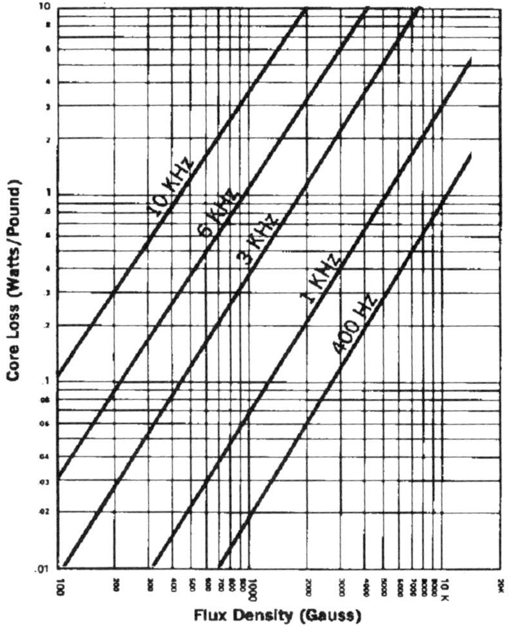

Figure 3. 19-Core loss curves for .002″ Alloy 48. From Magnetics (1995)

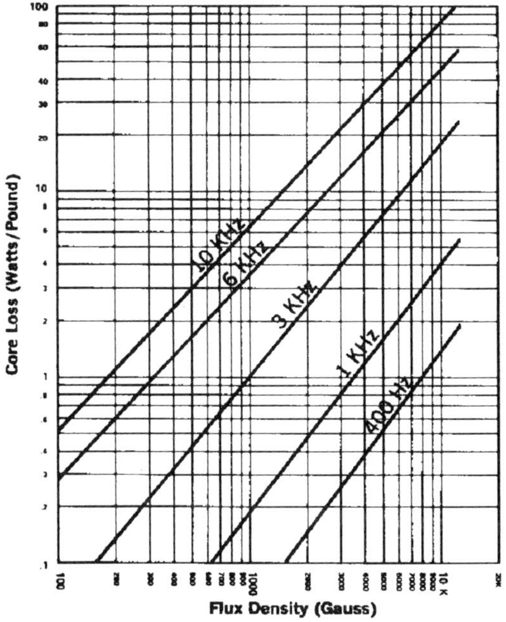

Figure 3.20-Core loss curves for .002″ Orthonol. From Magnetics (1995)

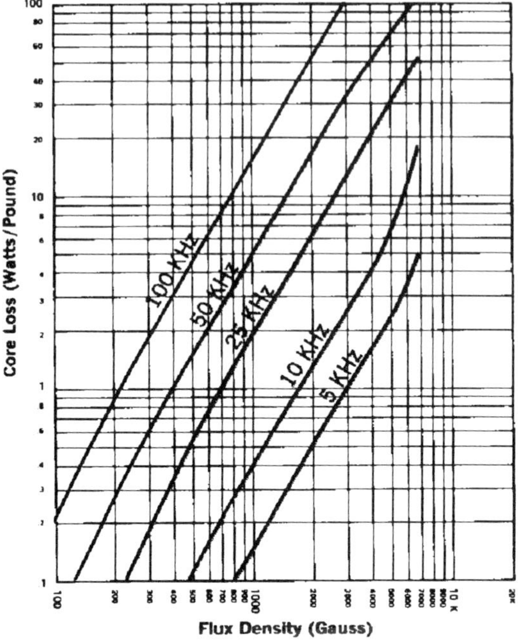

Figure 3.21-Core loss curves for .001″ square Permalloy 80.From Magnetics (1995)

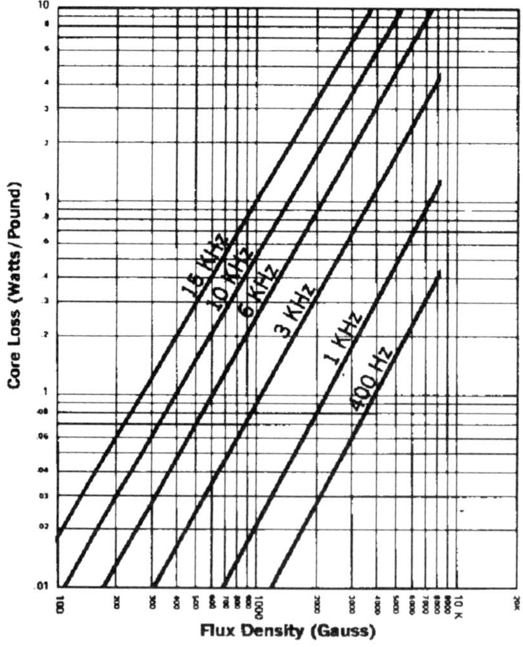

Figure 3.22-Core loss curves for .001″ Supermalloy .From Magnetics (1995)

ferromagnetic amorphous materials were made culminating in 1969 with the resulting continuous casting of the Metglas® alloys by Allied in the U.S. Over the next 20 years or so, a large number of amorphous magnetic alloys have been developed by Allied in production equipment and in fairly wide widths. The general composition of these alloys is $M_{80}B_{20}$ in which M is a magnetic metal such as Fe, Ni or Co and B is a non-metal such as B or a combination of non metals including B, Si, C, and P. The composition of the alloys can be tailored to accentuate one or more of the magnetic parameters such as permeability, saturation, squareness and low magnetostriction. Even in a single composition, the material can also be processed for high or low frequency applications.

These alloys are listed in Table 3.2 with their major magnetic properties. The best combination of characteristics for applications such as 60 Hz. distribution transformers, power transformers and motors would be the 2605TCA because of its low losses at 60 Hz, its high D.C. permeability and its moderately high saturation. These figures are measured on samples heat treated in a longitudinal magnetic field (See later under processing) but as such have only 30% of the losses if an M-2 oriented silicon steel. For airborne applications at 400 cycles where weight and volume considerations are critical, a higher saturation alloy, 2605CO containing cobalt might be better suited. The comparison of two 2605 alloys versus 4 mil (very thin)3% silicon-iron is shown in Figure 3.23.

3.3.1-Advantages of Amorphous Magnetic Metals

The distinguishing features of the iron-based materials that are responsible for the low losses are;

1. The high resistivity-about 137 μΩ–cm. Compared with about 47 μΩ–cm for 3% silicon iron.
2. The thin gage (about 1mil) compared to conventional 12 mil FeSi or the newer 7 or 9 mil materials
3. High saturation compared to other Ni or Co based materials

3.3.2-Disadvantages of Iron-based Amorphous Alloys

1. The saturation is not as high as iron or silicon iron. This may be due to the dilution of the magnetic iron atoms with the 20% non-metal atoms.

Table 3.2- -Listing of Various Amorphous Metal Magnetic Materials)

		As cast			*Annealed*		
Alloy	*Shape*	H_c (A/m)	M_r/M_s	μ_{max} (10^3)	H_c (A/m)	M_r/M_s	μ_{max} (10^3)
Metglas #2605 $Fe_{80}B_{20}$	Toroid	6.4	0.51	100	3.2	0.77	300
Metglas #2826 $Fe_{40}Ni_{40}P_{14}B_6$	Toroid	4.8	0.45	58	1.6	0.71	275
Metglas #2826 $Fe_{29}Ni_{44}P_{14}B_6Si_2$	Toroid	4.6	0.54	46	0.88	0.70	310
$Fe_{4.7}Co_{70.3}Si_{15}B_{10}$	Strip	1.04	0.36	190	0.48	0.63	700
$(Fe_{.8}Ni_{.2})_{78}Si_8B_{14}$	Strip	1.44	0.41	300	0.48	0.95	2000
Metglas #2615 $Fe_{80}P_{16}C_3B$	Toroid	4.96	0.4	96	4.0	0.42	130

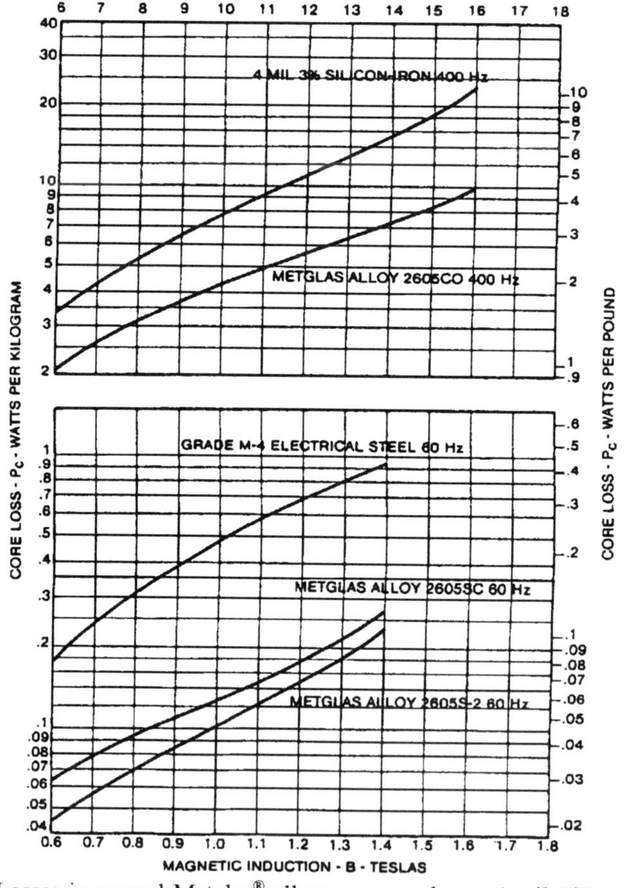

Figure 3.23- Losses in several Metglas® alloys compared to a 4 mil SiFe alloy. From Jiles (1991)

2. There is high magnetostriction in the iron-based alloys. This would tend to produce vibration and higher losses.
3. The amorphous alloys have an unusual combination for soft magnetic alloys in they are magnetically soft but mechanically very hard like a glass usually is. This hardness makes the amorphous alloys great for wear-resistant applications such as in recording heads (See later) but wreaks havoc when it comes to slitting, punching or other mechanical working. As a result, most amorphous core configurations are toroids. Even in a toroid, the brittleness creates difficulty in winding cores with too small an inside radius because of cracking and stresses.

3.3.3-Processing of Amorphous Metals

The constituent raw materials are placed in a induction melting apparatus enclosed in a container that can be pressurized. After melting, the molten glass is extruded under pressure through a narrow slit onto a rapidly turning water-cooled wheel such that the molten material solidifies as a glass by cooling at a rate of about one million degrees per sec. This cooling is so fast that crystallization cannot be accomplished before solidification. Of course, the chemistry is such that the crystallization time was comparatively slow. After casting and forming in a core, the material must be heat treated to bring out the best magnetic properties. In some very large magnetic structures such as generators or large transformers, the material may be used unannealed. Under the right heat treating conditions, the iron-based materials are especially useful at low frequencies in the range we have been discussing. With a different heat treatment, the same materials can be used in higher frequency magnetic applications. Since the material is extruded through a narrow slit, it is cast in the final gage thickness that is about 25 microns. The attractive feature of this method is that no further rolling is necessary. For other alloys, rolling to this thickness could be quite expensive. New manufacturing processes have permitted the production of moderately wide sheet material.

3.3.4-Amorphous Metals High-Perm-High Frequency

We have spoken of the iron-based amorphous Nickel-iron alloys for low-frequency, high power applications in Chapter 6 and the Ni-based amorphous alloys in Chapter 8. In this section, we will deal with the nickel and Co based amorphous alloys for much the same applications of the crystalline nickel-iron alloys discussed in this chapter. These are the applications for high permeability and high frequency operation. The first commercial amorphous alloy offered for sale was Allied Metglas® 2826. As in the case of the iron based 2605 alloys, the number indicates the major elements in the composition. Thus 28 stands for nickel(atomic number 28) and the 26 stands for iron (atomic number 26). The composition was $Fe_{40}Ni_{40}P_{14}B_6$. It has since been replaced with a better alloy, Metglas® 2826MB ($Fe_{40}Ni_{38}Mo_4B_{18}$). It most closely resembles the crystalline NiFe alloys. It has a μ_{max} of 800,000 after annealing. It has an improved saturation magnetostriction of 12 ppm. The resistivity is higher than the crystalline alloys at 138 μΩ-cm. And the Curie temperature is 353°C. It can be used in high frequency transformers and field sensors and shielding. It can be magnetically annealed for either square or flat hysteresis loops. There are two cobalt-based materials produced by Allied, Metglas® 2705M and Metglas® 2714A. Both have essentially zero magnetostriction. The iron-based amorphous alloys had high magnetostrictions. The 2705 material has a relatively high μ_{max} at 600,000 but it has another important property related to that. The permeability as cast (no anneal) is still a moderately high 290,000. This is a great saving for shield fabricators. This alloy is useful in high frequency cores and for magnetic sensors. It has a saturation of 7700 Gauss. The 2714A alloy has the highest μ_{max} of any amorphous alloys at 1,000,000. The saturation magnetostriction is much less than 1 ppm. Its resistivity is 142 μΩ-cm but its Curie point is only 225°C. The saturation is 5700 Gauss. It has extremely low core loss and excellent corrosion resistance. It can be used in high frequency transformers, high-sensitivity matching transformers, ultra-sensitive current transformers and in sensor applications. With all these excellent properties, it is not surprising that this is the most expensive amorphous alloy. As in the other amorphous alloys, the thickness is set at 20-30 microns (about .001″) dictated by the strip casting process. It has additional properties for recording head material because of its great hardness. However, its high brittleness must also be considered. The hysteresis loops for the 3 materials mentioned are found in Figures 3.24-3.26. The impedance permeability and core loss curves for the 2714A material are given in Figures 3.27-3.28. Vacuumschmelze also produces several Ni and Co based amorphous alloys. Vitrovac4040 is a FeNi based alloy that has a μ_{max} at 50 Hz. of 250,000, an H_c of 4 mA/cm (.005 Oersted) a saturation of 8700 Gauss, and a saturation magnetostriction of 1.35 ppm. The two Co based alloys are essentially zero magnetostriction. Vitrovac 6025 has a high μ_{max}, a saturation of 5500 Gauss,a resistivity of 135 μΩ-cm and a saturation magnetostriction of less than .3ppm. Vitovac 6030 has a μ_{max} of 300,000, λ_s = .3 but B_s is 8000 Gauss. The resistivity is 135 μΩ-cm . This alloy may be used in low loss inductive components especially at high frequency and also for magnetic sensors(anti-theft) and for recording heads.

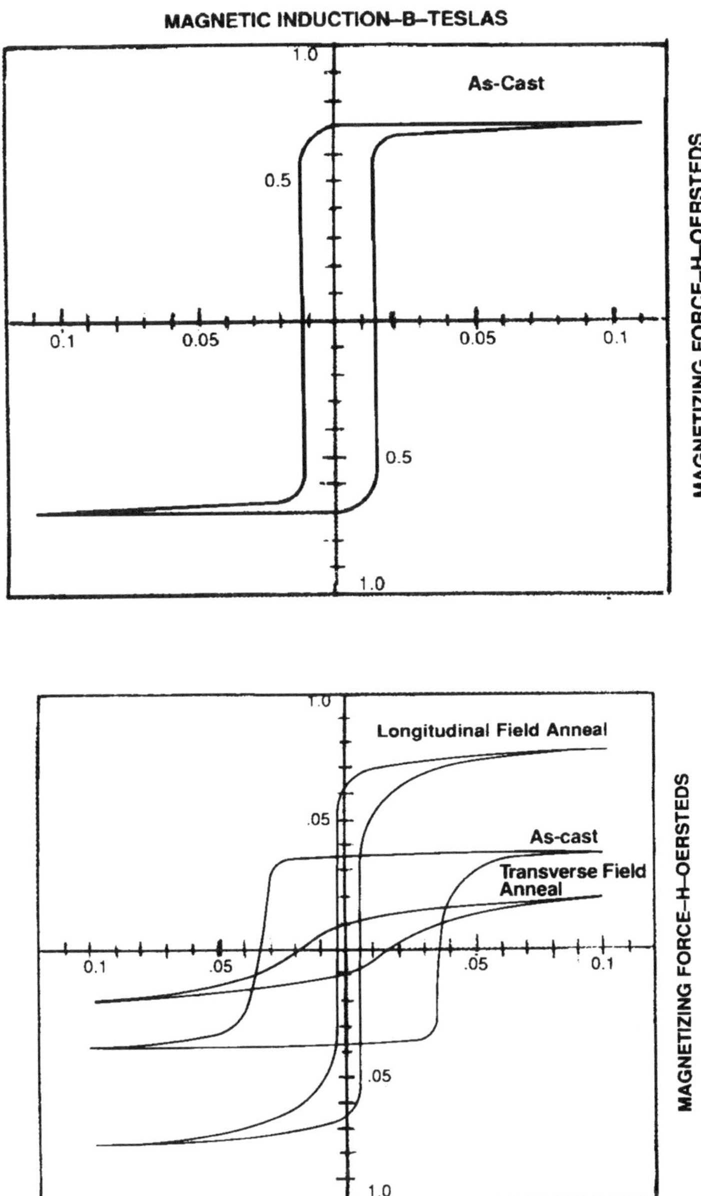

Figure 3.24- Hysteresis loops with various anneals for Metglas ® 2826MBFrom Allied (1995)

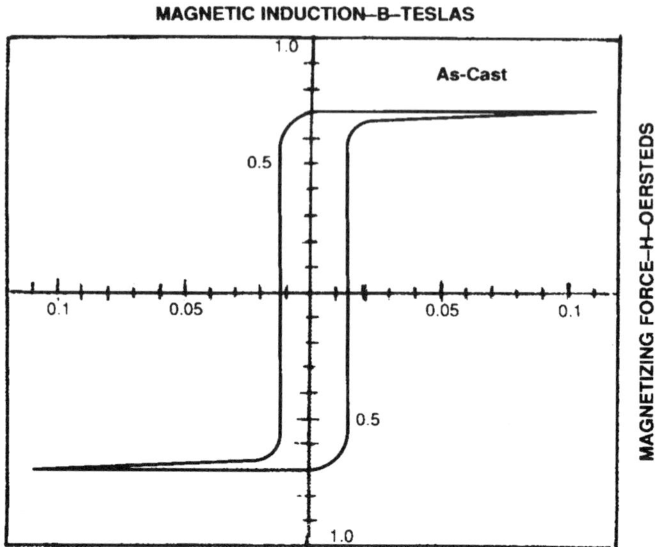

Figure 3.25-Hysteresis loop for Metglas ® 2705 From Allied (1995)

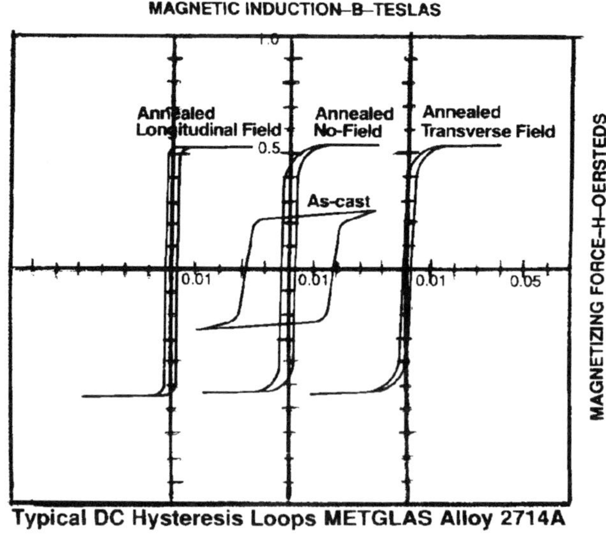

Figure 3.26- Hysteresis loops with various anneals for Metglas ® 2714A. From Allied (1995)

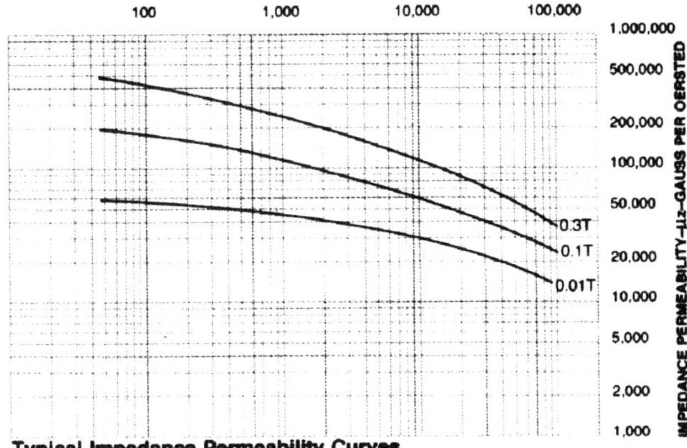

Figure 3.27- Permeability versus frequency losses for Metglas ® 2714A. From Allied (1995)

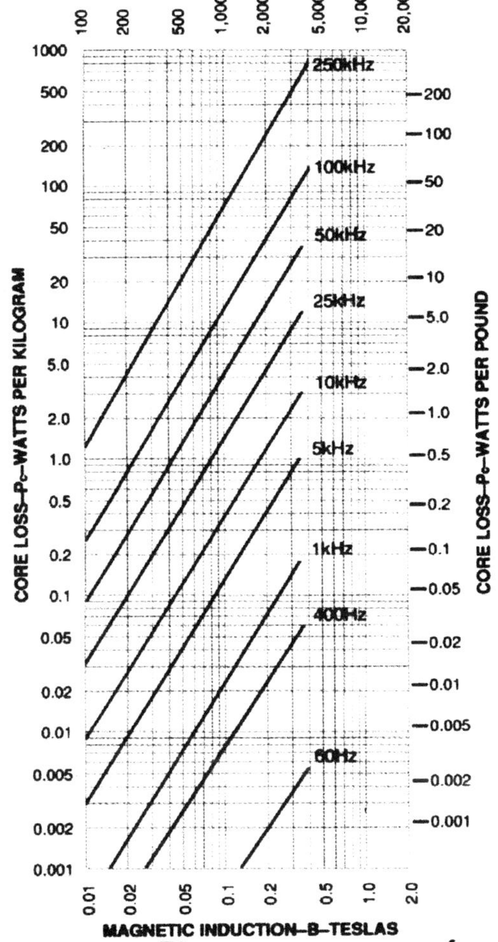

Figure 3.28- Core losses for Metglas ™ 2714A.No field anneal. From Allied (1995)

3.4-NANOCRYSTALLINE MATERIALS

The most recent of the soft magnetic metal materials are the nanocrystalline materials. They do have the relatively high saturation required for low-frequency power applications. However, the main application for these materials appears to be in the higher frequency, higher permeability area. First, the brittleness of the material (as it is in the amorphous metal material) dictates that toroids are the only practical shapes. Second, it is doubtful that the cost of the material can compete with the other power magnetic materials mentioned earlier in this chapter. The nanocrystalline materials will be dealt with in later chapters for other applications.

3.4.1-Nanocrystalline Materials for High Frequency Power

The iron-based nanocrystalline materials are the newest ones and have the potential of being the next generation of magnetic alloys. They are characterized by a very small grain size which lends itself to high frequency operation. The saturation and permeability for low magnetostriction materials are given in Figure 3.29 The first reported material with excellent magnetic properties was by Yoshizawa (1988). It came as an outgrowth of the amorphous metals studies. While earlier workers had tried to reduce the high magnetostriction of the partially crystallized iron based amorphous alloys by the use Nb, Mo or Cr, the properties were not good. Crystallization of the amorphous materials by heat treatment above the crystallization temperature, produced precipitates such as iron borides. The important contribution of Yoshizawa was the use of copper that prevented this precipitation effect. This permitted the use of the Nb to lower the magneostriction without harmful effects. The formula for his material that was called Finemet™ by Hitachi Metals was $Fe_{73.}5Cu_1Nb_3$ $Si_{13.5}B_9$. The amorphous metals were made by the single roller method in widths of 5mm. and thickness of 15-20μm. There were made into while the CuNb region remains amorphous (higher crystallization tempera cores with an OD of 19 mm, an ID of 15 mm. and annealed at 380-650°C. The magnetic properties as a function of Cu content are given in Figure 3.30. The saturation is high at 1.2-1.3 T. The permeability at 1 Khz. Is 100,000, the H_c was .5A/m. The Cu markedly improved the properties of the FeNbSiB alloy with respect to all magnetic properties but the saturation. The crystallization temperature was 488°C. For the 1%- Cu alloy, the copper has the effect of nucleating the bcc-Fe solid solution and preventing the precipitation of the borides. In the alloy, there are Fe rich regions and Cu-Nb rich regions. The iron rich region crystallizes because of the lower crystallization temperature.

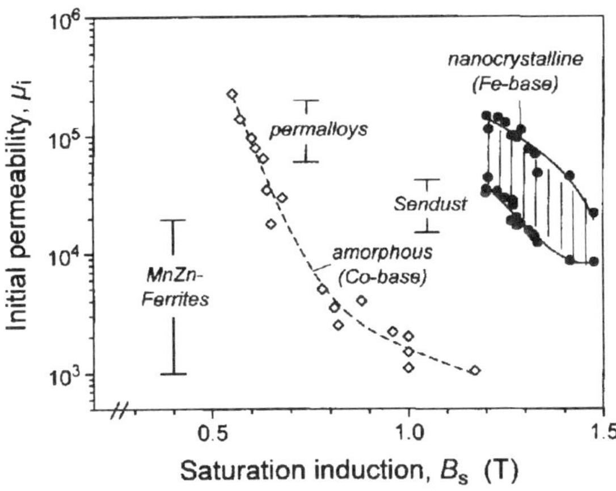

Figure 3.29 Saturation and permeability of some low magnetostriction materials. Reprinted from Herzer, Handbook of Magnetic Materials Vol. 10, ©1997), p. 454, with permission from Elsevier Science.

The iron -rich region cannot grow into the Cu-Nb region so the grain size remains small and the texture is random. Therefore the saturation magnetostriction is low (2 x 10^{-6})compared to the iron based amorphous alloys. In a later paper (Yoshizawa 1989) was able to magnetically anneal these materials and obtain square or flat loops by longitudinal or transverse anneal respectively. The square loop material was inferior to other materials but the flat loop (transverse) material showed good properties to 150 KHz. and the permeability was 30,000. Results are shown in Figure 3.31. Hitachi offers standard tape-wound toroids of the Finemet material. Late Suzuki (1990) developed another nanocrystalline alloy system based on Fe-Zr-Nb-B. This was later expanded and reported by Makino(1997). Permeabilities up to 160,000 were obtained and in other alloys, saturations to 1.7 T (17,000 Gauss). The material is called Nanoperm . Most of the materials described here are experimental and laboratory results. Aside from the Hitachi tape cores of the Finemet material. Vacuumschmelze offers a commercial product called Vitroperm which is a low magnetostriction alloy with high permeability and exceptionally low core losses. The saturation is 12,000 Gauss. Herzer (1989) has pointed out the anomaly of the nancrystalline materials with respect to the influence of grain size on the coercive force, H_c, In conventional crystalline materials H_c varies inversely with the grain diameter (1/D). Thus, the larger the grain size, the lower the coercive force. In the nanocrystalline materials, the coercive force varies as the sixth power of the grain size (D^6) Therefore, the coercive force decreases strongly as the grain size decreases. This is shown in Figure 3.32. Herzer explains this by noting that the small size of the grains are smaller than the ferromagnetic exchange length and thus, the magnetization cannot follow the easy axis of magnetization. The texture is then

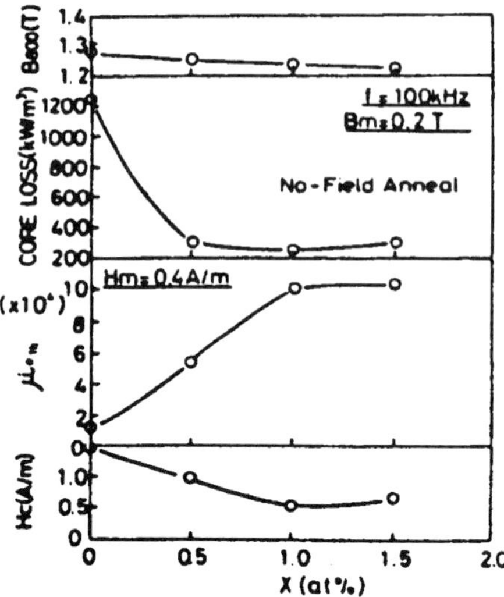

Figure 3.30-Magnetic properties of nanocrystalline material as a function of Cu content. From Yoshizawa (1988)

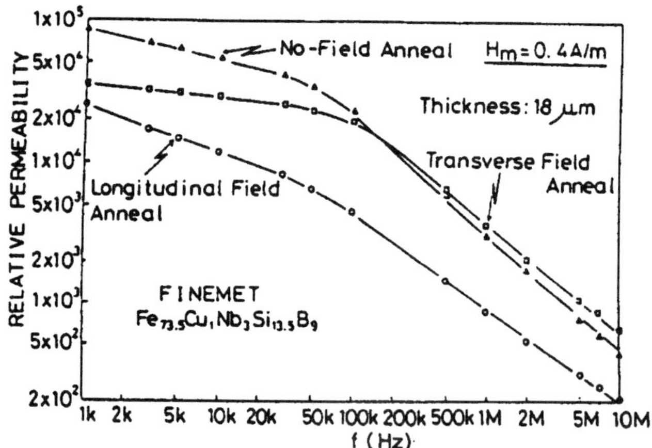

Figure 3.31- Permeability versus Frequency for Finemet material with different anneals From Yoshizawa (1989)

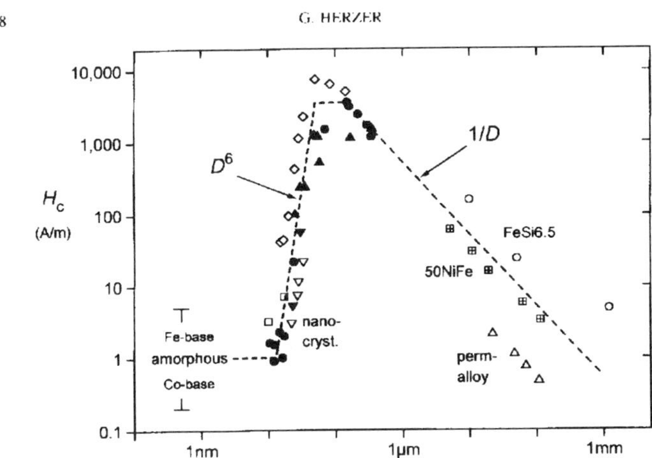

Figure 3.32-Variation of coercive forces of various materials with grain size. Reprinted from Herzer, Handbook of Magnetic Materials Vol. 10, ©1997), p418, with permission from Elsevier Science.

G. HERZER

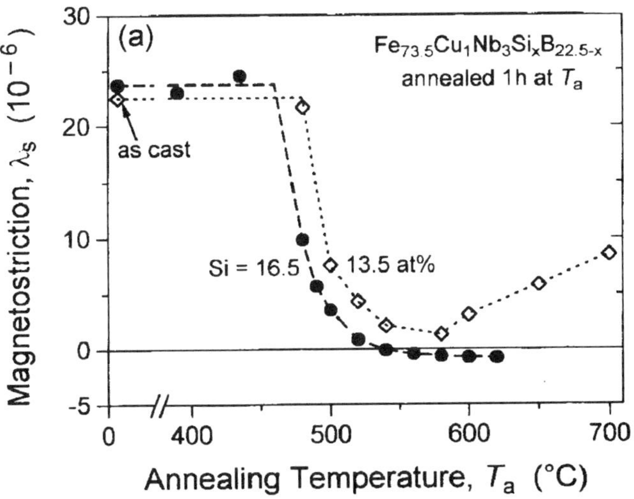

Figure 3.33-Dependence of magnetostriction on annealing temperature for a nanocrystalline material with varying Si content Reprinted from Herzer, Handbook of Magnetic Materials Vol. 10, ©1997), p. 444, with permission from Elsevier Science.

random and the H_c is low and the permeability is high. In addition to the anisotropy, the magnetostriction is also low. The change in magnetostriction with annealing temperature is shown in Figure 3.33. The magnetostriction is also a function of Si content.. The permeability and loss factor for the nanocrystalline material as compared to the MnZn ferrite or Co based amorphous material is shown in Figure 3.34. It shows improved properties over the ferrite and similar properties to the amorphous material. In addition, the saturation of the nanocrystalline material is higher and also has better thermal stability than the other two. Hilzinger points out that in addition to the switched mode power application (to be discussed in a later chapter), there are low level uses as in signal transformers, common mode chokes, pulse transformers, current transformers and ground fault interrupter cores.

This new generation of magnetic materials is still in its infancy and it is hard to predict the extent that it will improve our use of magnetic materials.

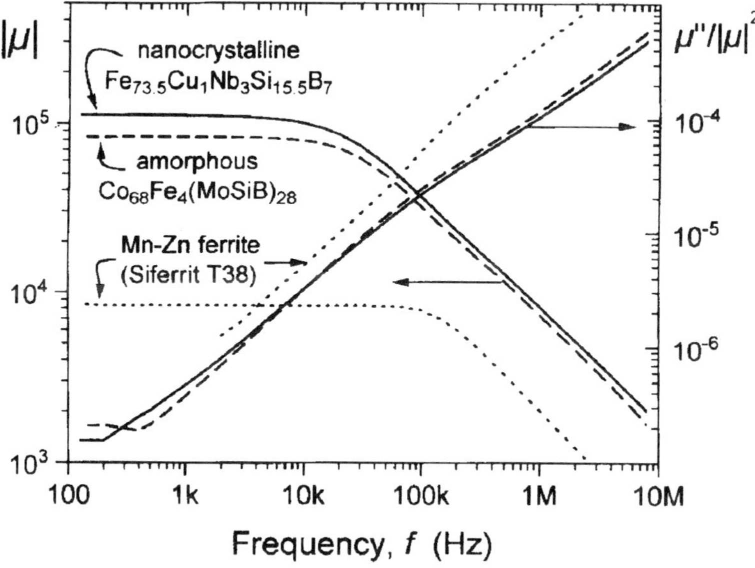

Figure 3.34-Permeability versus frequency for various materials. Reprinted from Herzer, Handbook of Magnetic Materials Vol. 10, ©1997, p.455, with permission from Elsevier Science.

3.4.2-Amorphous-Nanocrystalline Materials for EMI Suppression

One of the earliest uses of the amorphous material described earlier was as a choke coil that Toshiba called the "Spike-killer". Presumably, only the cores are sold. The amorphous materials used are under license from Allied's Metglas® Division. Vacuumschmelze does market a Co-based amorphous material for EMI suppression applications. It is designated Vitrovac 6025 and is essentially a zero magnetostriction material. The characteristics are shown in Table 3.3. As an outgrowth of the amorphous materials, the iron-based nanocrystalline materials are the newest ones available and they have been used for EMI suppression. Their high permeabilities and low magnetostrictions made it very useful as a common-mode choke at relatively low frequencies. The earliest nanocrystalline soft magnetic alloy was made by Hitachi and names “Finemet”. The properties of the Finemet nanocrystalline material are given in Table 3.4. The saturation is high but not in the Si-Fe class. The permeability is high but not in the 80 permalloy class. The magnetostriction is low and about the same as the Co-based amorphous material. The resistivity is the same as the amorphous alloys but many orders of magnitude lower than ferrite. It is really the combination of most of the good attributes that make it attractive. High permeability materials without the high resistivities are useful for the inductive or permeability portion of the impedance and are limited to the lower frequencies.

Vacuumschmelze has two nanocrystalline materials, Vitroperm 500F and 800F that they recommend for EMI suppression(common-mode chokes) along with their Co-based amorphous zero-magnetostriction material Vitrovac 6025. The permeability of the 800F is somewhat higher than the 500F. A comparison of the Co-based amorphous, the nanocrystalline material and a MnZn ferrite is given in Figure 9.35 The permeability is higher and the loss factor is lower for the metallic material than for the other materials. The in-

Table 3.3 Properties of Amorphous and Nanocrystalline Materials for EMI Suppression Applications

				6025 F	500 F
Sättigungsinduktion /	saturation induction	Bs	[T]	0,55	1,2
Sättigungsmagnetostriktion /	saturation magnetostriction	λs	[ppm]	0,20	<0,5
Spez. elektr. Widerstand /	resistivity	ρ	[Wmm^2/m]	1,35	1,15
Curietemperatur /	curie temperature	T_c	[°C]	210	600
Banddicke /	tape width		[μm]	25	20
Stat. Koerzitivfeldstärke /	static coercive field strength	H_c	[A/cm]	0,003	0,005
Ummagnetisierungsverluste /	hysteresis losses	(P_{Fe} 100 kHz; 0,3 T)	[W/kg]	100	105
Dichte /	density	ρ	[g/cm^3]	7,70	7,35

From Vacuumschmelze ,1994

sertion damping curves (50 ohms) versus frequency for the Vitroperm 500F with different windings and that of a MnZn ferrite are in Figure 3.26

3.5-POWDER CORES FOR OUTPUT CHOKES, EMI AND PFC

The use of gapped ferrite cores as output chokes was discussed earlier. in Section 3.1.10 . Another series of good material choices for the application are the metal powder cores. Aside from output chokes, a recent and

Table 3.4- Properties of "FINEMET" Nanocrystalline Material

		FINEMET FT-1KM	Mn-Zn ferrite	Co-based amorphous	Fe-based amorphous
Initial magnetic permeability uri	10kHZ	>=50,000	5,300	90.000	4.500
	100kHz	16,300+/-30%	5,300	18,000	4,500
Maximumn flux density BMS*(T)		1.35	0.44	0.53	1.56
Coercive force Hc*(A/m)		1.3	8.0	0.32	5.0
Rectangular ratio Brms/Bms*		0.60	0.23	0.50	0.65
Core loss Pc**(kW/m^3)		350	1,200	300	2,200
Curie temperature Tc(C)		570	150	180	415
Saturation magnetostriction(x10^-6)		+2.3	-	-0	+27
Specific resistivity(ohm-m)		1.1x10^-6	0.20	1.3x10^-6	1.4x10^-6
Density ds(x10^3kg/m^3)		7.4	4.85	7.7	7.18

From Hitachi 1998

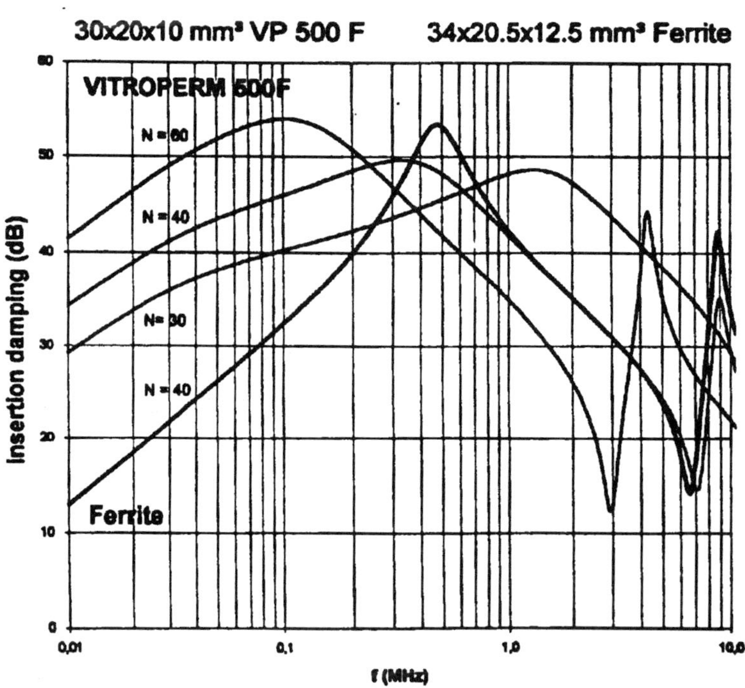

Figure 3.35-Insertion damping curves of a common-mode choke made of Vitroperm 500F for different numbers of turns compared to a choke made of ferrite.From Hilzinger , 1996,

related application is that of power factor correction (PFC) cores. For these applications, there are two main advantages of powder cores over ferrites. They are;

1. Their saturation flux densities are much higher (as much as 2-3 times higher)
2. The gap is a distributed one while that of ferrite is discreet leading to high gap losses.

A disadvantage of the powder core is the need for costly toroidal winding as they are mostly used as ring cores. Some metal powder E-cores are now available. Similar to the metal strip analog, the lowest loss, highest permeability material of the powder cores are the 80% Ni variety which allows for operation at higher frequencies (especially in thin gage). The 50% Ni alloy has twice the saturation of the 2-81 Moly Permalloy (MPP) material but it has higher losses and is used for lower frequencies. The iron powder cores have the highest saturation and lowest cost but the highest losses.

For EMI applications, while common-mode filters are mostly used in unbalanced circuits where the currents return to ground, the differential-mode filter is used primarily in balanced systems. Consequently, putting a ferrite toroid around both wires would not cause any flux change in the core and so not suppress the EMI. The solution in this case is to put suppressor cores on each of the wires. However, this means that the full ac (and D.C.) signals would pass through the suppressor core. While the common-mode ferrite core can be used in high current power filters, the differential-mode suppressor ferrite core is only used in low current power filters. With the differential-mode or in-line filters, the core losses could be a problem (except for D.C.), the main problem would be core saturation. To prevent this, a core with low permeability is needed. Either a gapped ferrite core or a powder core can be used.

3.5.1-Iron Powder Cores

The iron powder cores, unlike those listed for low level telecommunications applications are of the higher permeabilities (from 70-100 perm). They are usually of the hydrogen-reduced variety. Before discussing the magnetic properties of powder cores for EMI suppression, we must point out that, while permeability has been our criterion for EMI ferrite suppression ability, with powder cores, vendors do not specify impedance. Since the application of these cores involves high flux densities and often D.C. bias, the magnetic properties usually listed are;

1. Permeability versus Flux Density
2. Permeability versus DC Bias
3. Core Losses
4. Permeability versus Frequency
5. Permeability versus Temperature
6. D.C. Energy storage curves

Another important property for an application as a D.C. choke is the variation of energy stored versus D.C. current. The energy storage criterion is given by ½ LI^2. For the iron powder cores, the high saturation of about 20,000 Gauss is suited for this application. Curves displaying the variation of permeability with flux density and DC bias for iron powder cores are shown in Figures 3.36 and 3.37. Permeability versus frequency is shown in Figure 3.38.Core loss and Energy storage curves are shown in Figures 3.39-3.40.

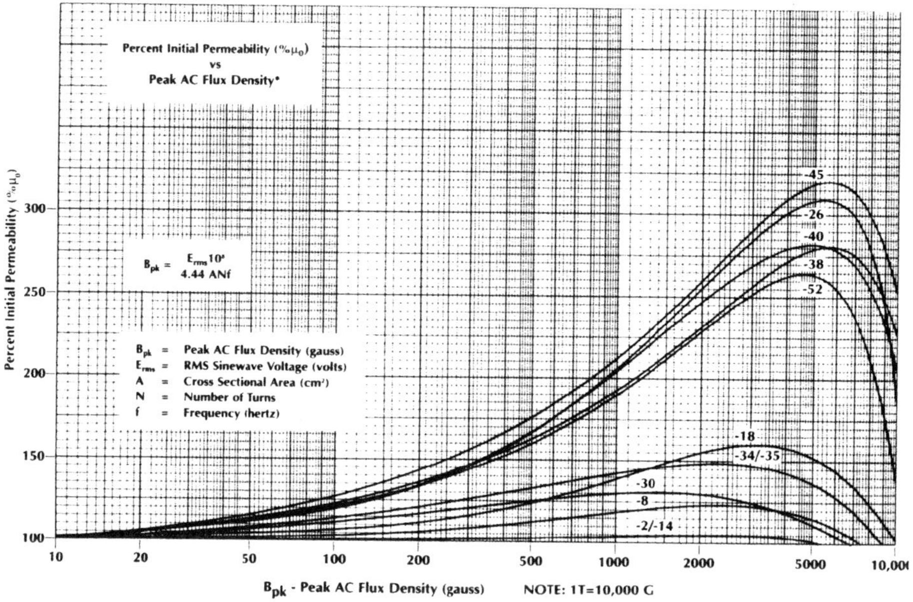

Figure 3.36-Permeability vs Flux Density for several permeabilities of iron powder cores (From Micrometals)

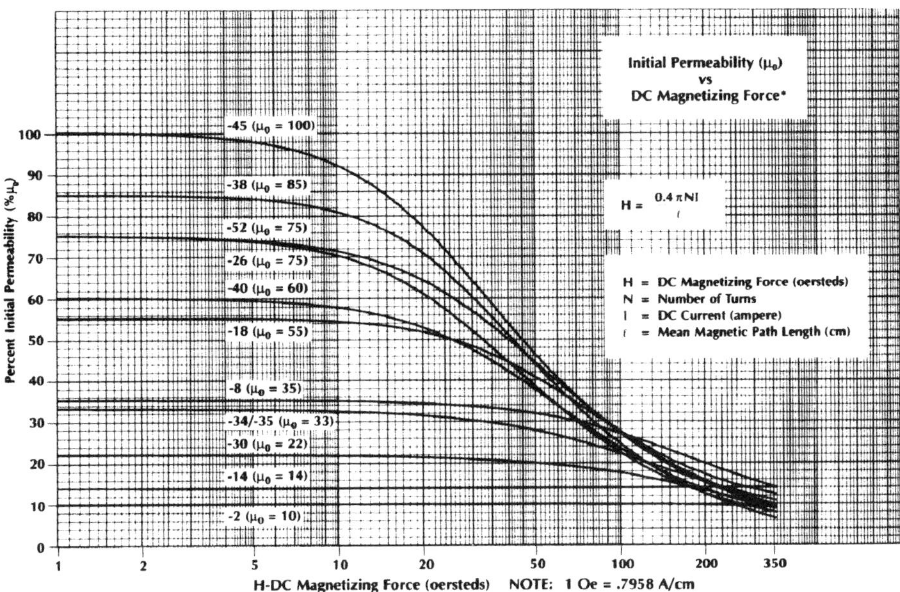

Figure 3.37-Permeability vs DC Bias for several permeabilities of iron powder cores (From Micrometals)

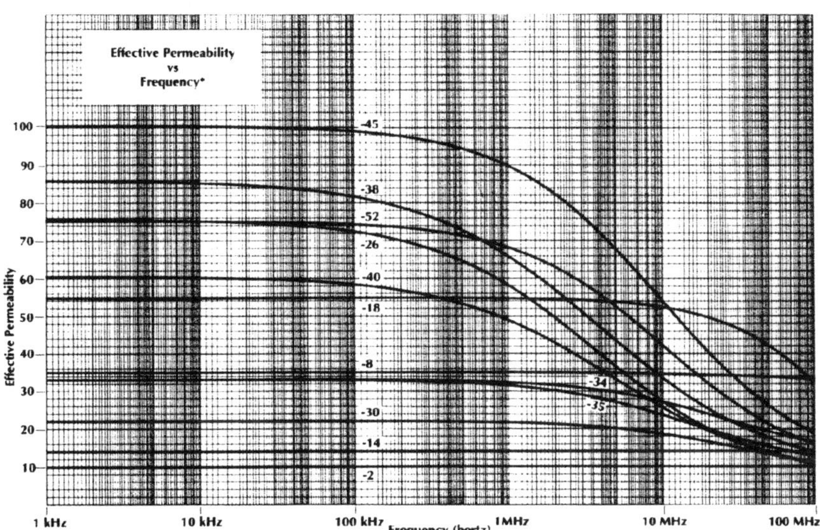

Figure 3.38- Permeability vs Frequency for several permeabilities of iron powder cores. (From Micrometals)

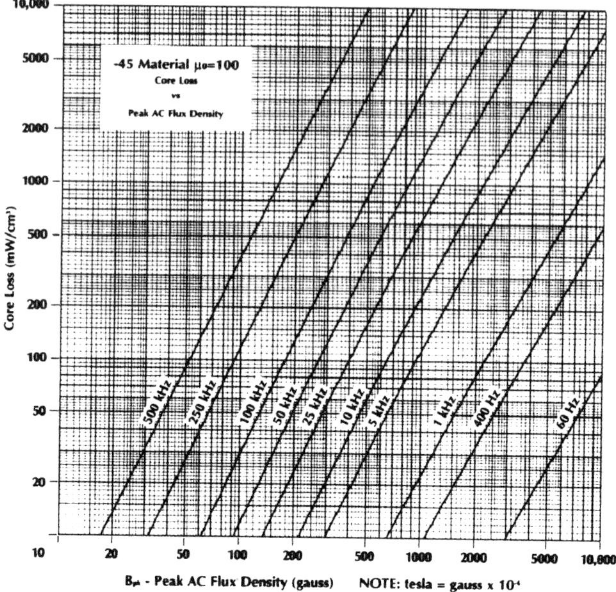

Figure 3.39-Core losses vs flux density and frequency for a 100 perm iron powder core. (From Micrometals

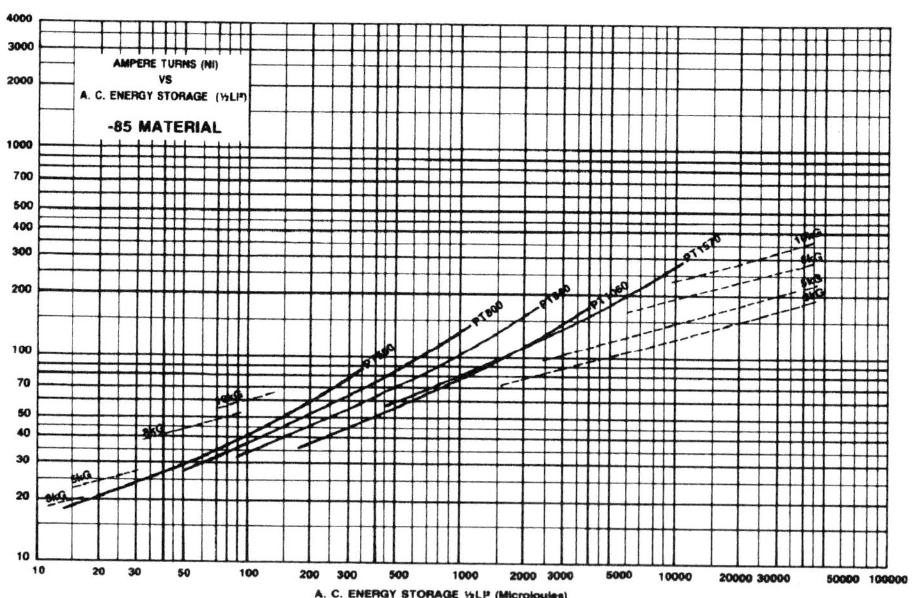

Figure 3.40- Energy storage curves for 85 perm iron powder cores. From Pyroferric

3.5.2-NiFe Powder Core Materials

Nickel iron powder cores come in two varieties, the 2-81 MPP (Molypermalloy Powder) and the 50-50 Hi Flux Cores.The NiFe powder cores listed as High Flux cores are different from the MPP cores listed in the chapter on low level applications of powder cores. These NiFe cores are 50% Nickel-50% Iron. The have about twice the saturation (15,000)of the MPP cores and thus are much better for the present application. Cores of this material are available in permeabilities of 200, 160, 147, 125, 60, 26 and 14. The variations of permeability with flux density and D.C. bias, frequency and temperature for different permeabilities of this material are shown given in Figures 3.41 and 3.42. Plots of permeability versus frequency and temperature are found in Figures 3.43 and 3.44.The core losses are given in Figure 3.45. As expected, the stability is inversely proportional to the permeability but the high frequency core losses are proportional to the frequency. Cost-wise the High Flux powder cores are more expensive than the iron powder cores, but somewhat less expensive than the MPP cores.

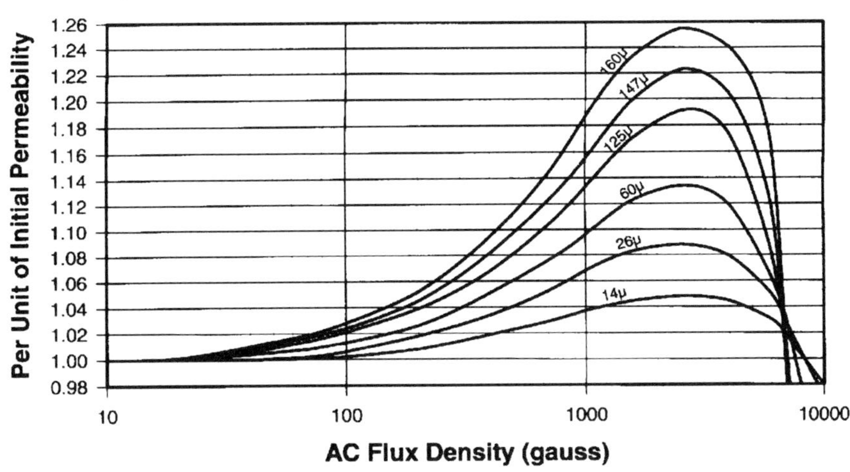

Figure 3.41-Permeability versus flux density for different permeability NiFe High Flux Powder Cores. From Magnetics 1998

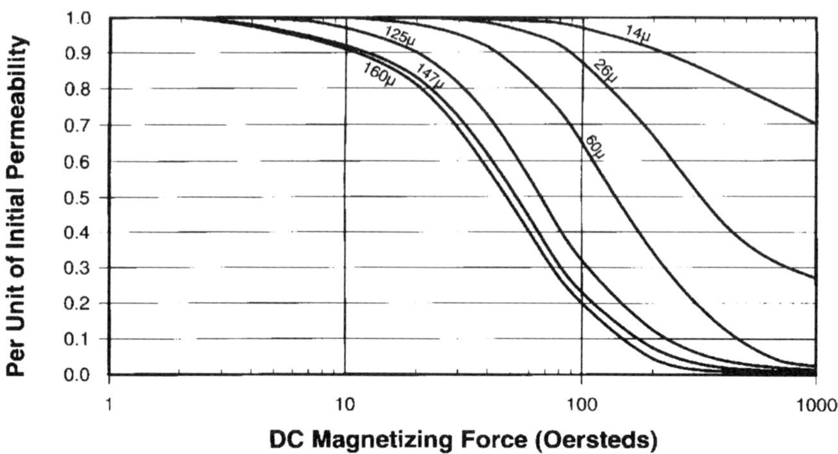

Figure 3.42-Permeability versus DC bias for different permeabilty NiFe High Flux powder cores. From Magnetics 1998

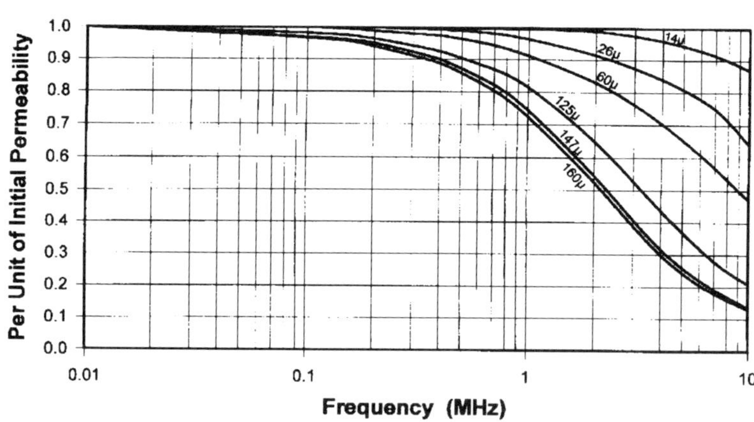

Figure 3.43 -Permeability vs frequency for different permeabilty NiFe high flux powder cores. From Magnetics 1998

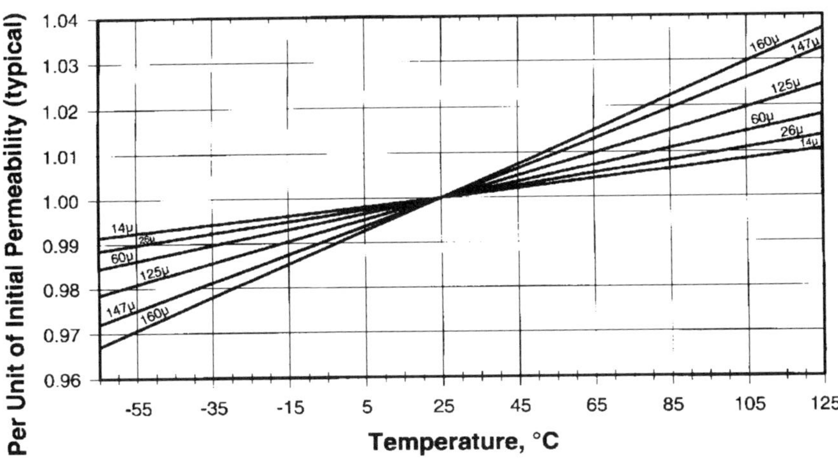

Figure 3.44-Permeability vs temperature for different permeabilty NiFe high flux powder cores. From Magnetics 1998

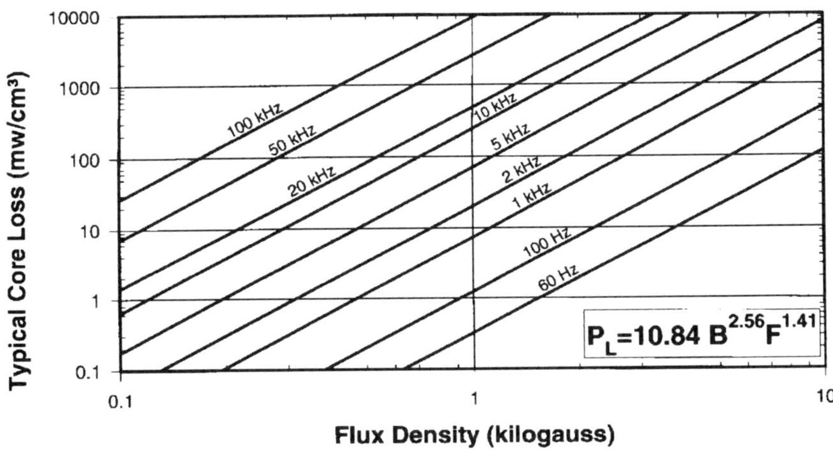

Figure 3.45-Core loss curves for different permeabilty NiFe high flux powder cores. From Magnetics 1998

3.5.3- Sendust Powder Material

The Sendust cores are marketed under the trade names of Kool-Mu and MSS materials. It is a new application of an old material having been described by Matsumoto in 1936 and patented in 1940 (Matsumoto 1940). Sendust is a ternary alloy containing about 6% aluminum and 9% silicon. Its attraction is that it is close to a zero anisotropy-zero magnetostriction material. Its brittleness and difficulty in producing it have limited its use in the past to recording head material due to its great hardness. When used in powder cores, its brittleness helps in the comminution process. The high saturation of this material (on the order of about 10,000 Gauss provides much more energy storage than MPP cores or gapped ferrites. The cores come in permeabilities of 60, 75, 90 and 125. Figures 3.46-3.49 show the permeability variations of different permeabilities of this material for flux density, D.C. bias, frequency and temperature. The core loss of the 125 perm material is given in Figure 3.50. The core losses are significantly lower than the iron powder cores. However, the Sendust cores are somewhat more expensive. Cores with O.D's from .140 inches to 2.25 inches are available.

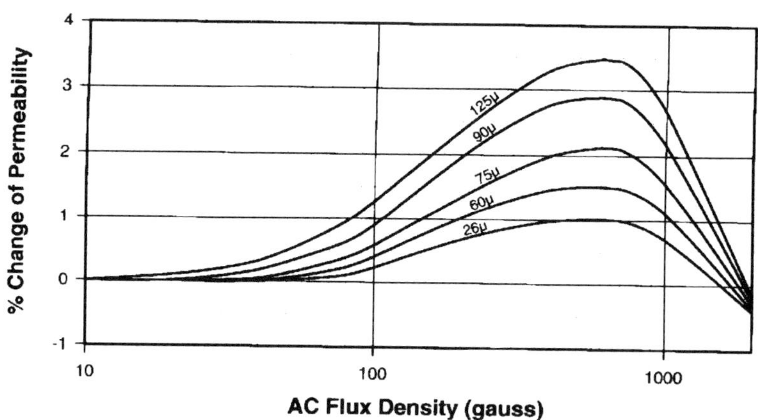

Figure 3.46- Permeability vs flux density for several different permeability Kool-Mu (Sendust) Cores . From Magnetics 1998

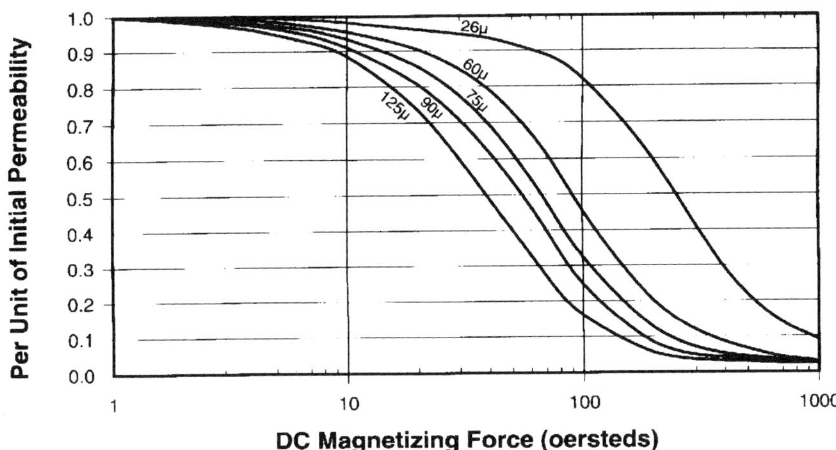

Figure 3.47- Permeability vs DC bias for several different permeability Kool-Mu (Sendust) cores . From Magnetics 1998

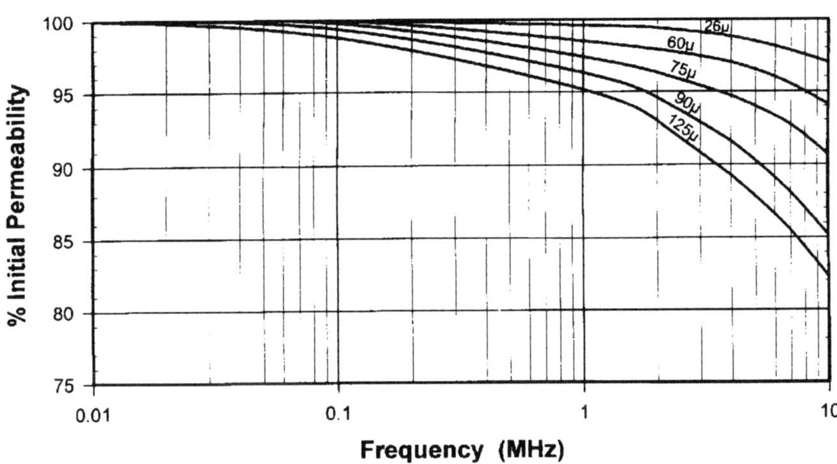

Figure 3.48- Permeability versus frequency for several different permeability Kool-Mu (Sendust) cores . From Magnetics 1998

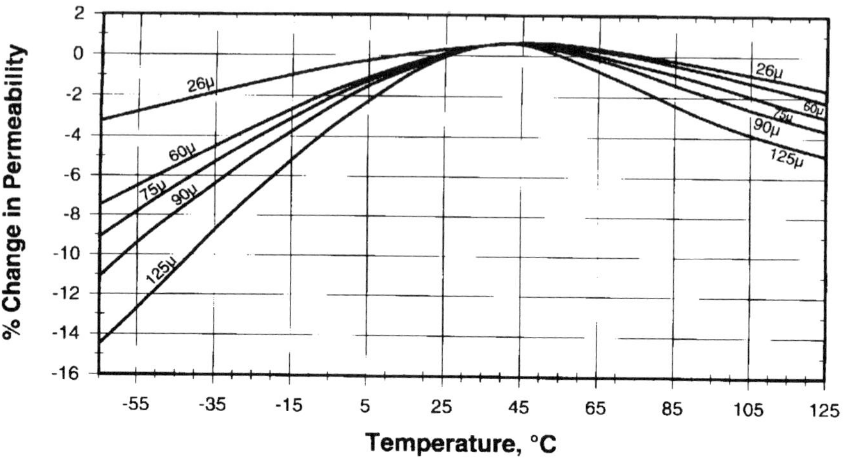

Figure 3.49- Permeability versus temperature for several different permeability Kool-Mu (Sendust) cores . From Magnetics 1998

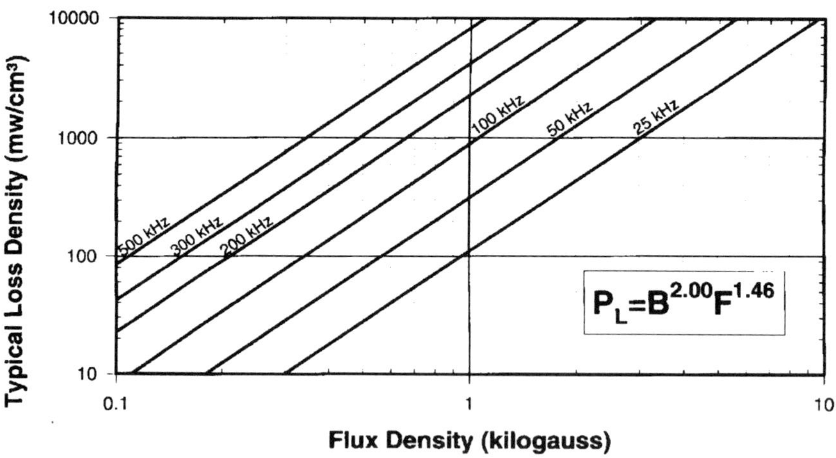

Figure 3.50- Core loss Curves for 125 perm Kool-Mu (Sendust) powder cores

References

Buthker,C.(1986) and Harper, D.J., Transactions HFPC, 1986,186

Bozorth, R.M. (1951) Ferromagnetism, Van Nostrand New York

Fair-Rite (1996) Fair-Rite Soft Ferrites, 13th Ed. Fair-Rite Products Corp. One Commercial Row, Wallkill, NY 12589

Goldman, A. (1985), Advances in Ceramics 15, Proc 4th ICF, p.421

Goldman, A. (1995), J. Mat. Eng. And Performance 4, 395

Herzer, G. (1997) Handbook of Magnetic Materials, Vol. 10, Elsevier Science B.V. Amsterdam, 418,444,454,455

Hitachi (1998) FINEMET FT-1KM-KN Series Core Page on Internet Product Guide

Hilzinger, H.R. (1996)Soft Magnetic Materials '96, Feb.26-28, 1996, San Francisco, Gorham-Intertech Consulting, 411 U.S. Route One, Portland ME, 04105Magnetics (1987) Ferrite Core Catalog, Magnetic Div.,Spang and Co, Butler, PA 16001

Honeywell (2000). Metglas® Technical Bulletin, Metglas® Products, 6 Eastman Rd. Parsippany NJ 07054

Magnetics (1995) Tape-Wound Cores Design Manual, TWC-400, Magnetics, Division of Spang and Co., Butler, PA 16001

Magnetics (1998) Powder Cores MPP Cores for Filter and Inductor Applications, Magnetics, Div. of Spang and Co. Butler, PA 16001

Magnetics (2000) Ferrite Core Catalog, FC601, Magnetic Div., Spang and Co, Butler, PA 16001

Makino, A. (1997),Hatanai, T., Naito, Y. Bitoh, T.,Inoue, A., and Masumoto, T., IEEE Trans. Mag. **MAG33**, 3793

Micrometals (1990) Micrometals Iron Powder Cores, EMI and Power Filters Micrometals, 1190 N. HawkCircle, Anaheim ,CA, 92807

MMPA (!996) Soft Ferrites, A User's Guide SFG-96

Parker, C. (1994) Presented at MMPA Soft Ferrite Users Conference, Feb. 24-25, 1994, Rosemont ,IL

Pyroferric(1984) Toroidal Cores for EMI and Power Filters, Pyroferric International, 200 Madison St., Toledo, IL 62468

Roess, E.(1982), Transactions on Magnetics MAG18,#6,Nov.1982

Smit,J.(1954) and Wijn,H.P.J. Advances in Electronics and Electron Physics, 6,69

Snelling E.(1988) Soft Ferrites, Properties and Applications Butterworths, London

Vacuumschmelze (1995) Vitrovac 500F-Vitroperm 6025, PK-004, Vacuumschmelze GMBH, Hanau, Germany

Yoshizawa, Y (1988) Oguma, S. and Yamaguchi, K.,J. Appl. Phys.,**64**, 6044

Yoshizawa, Y (1989) and Yamaguchi, K., IEEE Trans. Mag. **MAG25**, 3324

Chapter 4
CORE SHAPES FOR POWER ELECTRONICS

INTRODUCTION

In the previous chapter, the inherent material properties of components for power electronics were examined. In most cases these properties were measured on toroids because their magnetic cross-sectional area is constant and they have an uninterrupted magnetic path. This makes for ease of interpretation of the measurements. However, while toroids are still used in some applications, designers of magnetic circuits (including those for power electronics) find it more practical to rely on many other shapes for technical and economic reasons. Because the shape of the component influences the performance of the device, modified component parameters including material and shape considerations must be developed. This chapter will list the possibilities of core shapes used in power electronics. In addition, several new changes in the overall height to cross-section brought about by mounting on PC boards will be discussed. Since, very often the magnetic component is the largest on the board, the shape of the component takes on much more importance.

4.1-FERRITE CORE SHAPES

Ferrite cores possess one advantage over other magnetic materials in that they come in a large variety of shapes. This feature is made possible by the part-forming process in which the ferrite powder is pressed in a die before sintering to final dimensions. The die can be complex as long as the pressed part can be ejected from the die. Some parts such as round-leg E-cores must be pressed with legs up which creates a need for a minor adjustment. A variety of ferrite shapes for power applications are shown in Figure 4.1

4.1.1 Pot Cores

Pot cores are sometimes used ungapped in power applications with a solid center post since there is no need for the adjustor found in telecommunication applications. The shielding to protect a low-level telecommunication signal in LC circuits is not necessary. There may be some advantage to the shielding in that it does provide the lowest leakage inductance. Besides cost, another drawback to pot cores is the difficulty of bringing out heavy leads to carry the high currents. The closed structure also makes it difficult for heat

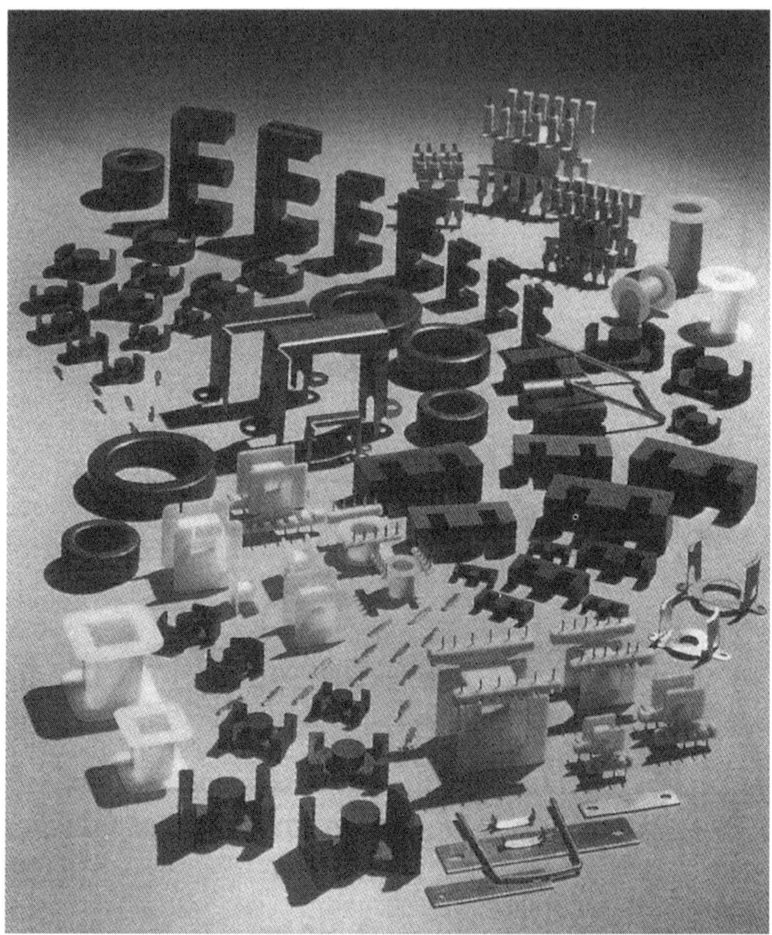

Figure 4.1-A variety of ferrite core shapes for power applications

from the windings to escape. Since pot core dimensions all follow IEC standards, there is interchangeability between manufacturers.

4.1.2-Double Slab Cores

In slab-sided solid center pot cores, a section of the core has been cut off on each side parallel to the axis of the center post. This opens the core considerably. These large spaces accommodate large wires and allow heat to be removed. In some respects, these cores resemble E-cores with rounded legs. See Figure 4.2

Figure 4.2- Double-slab pot cores. The solid center-post cores are useful for power applications .

4.1.3-RM Cores and PM Cores

RM cores (See Figure 4.3) were originally developed for low power, telecommunications applications because of the improved packing density. They have since been made in larger sizes without the center hole. Their large wire slots are an advantage while still maintaining some shielding PM cores are large RM-shaped cores specifically for power applications. Zenger(1984) feels that the geometry and self-shielding of RM cores make them useful at high frequencies. Roess (1986) points out that the stray field from an E-42 core is 5 times higher than that of an RM core. With the trend towards increased operating frequencies, he feels that there may be a backswing to the RM cores in mains (line) applications. Since that time, the use of RM cores for power applications has grown significantly.Low-profile RM cores are available in the RM4, RM5, RM6, RM7, RM8, Rm10, RM12 and RM14 sizes . Surface mount bobbins are available in RM 4 Low Profile, RM5, RM6, and RM6LP. For power non-linear choke cores, Siemens offers special RM8 to

RM 14 cores with tapered center posts. PM (Pot-core Module) cores are used for transformers handling high powers, such as in pulse power transformers in radar transmitters, antenna matching networks, machine control systems, and energy-storage chokes in SMPS equipment. It offers a wide flux area with a minimum of turns, low leakage and stray capacitance. Because of the weight of these pot cores, they may not be suitable for mounting on PC boards.

The numbering system of the RM cores is based on the grid system for holes on printed circuit boards. There are 10 grids to an inch (25.4mm) The RM number corresponds to the number of grids that a side of the square that contains the core. Thus an RM4 core would fit in an are of 4X4 grids (0.4X 0.4 inches) or about 10 x 10 mm.

Figure 4.3- RM Cores (hexagonal) and EP Cores (square). Courtesy of Magnetics, Division of Spang and Co.,Butler,Pa.)

4.1.4-E Cores

These cores are the most common variety used in power transformer applications. As such they are used ungapped. There are some variations that we shall discuss here. Their usefulness is based on their simplicity. Initially, E- cores were made from metal laminations and the early ferrite E cores were made to the same dimensions and were called lamination sizes. However, as the ferrite industry matured, E core designs especially useful for power ferrite applications were developed.(Figure 4.4). Many standard E-cores have bobbins that permit horizontal mounting. Some of the smaller sizes also are available in surface mount design with gull-wing terminals.

Figure 4.4- Ferrite E cores for power applications (Courtesy of Magnetics, Division of Spang and Co. Butler, Pa.)

4.1.5-E-C Cores

E-C cores are a modification of the simple E core. The center post is round similar to a pot core and since round center bobbins wind easier and are more compact than square center bobbins, this is an advantage. The length of a turn on the round bobbin is 11 percent shorter than the square bobbin that means lower winding losses. The legs of these cores have grooves to accommodate mounting bolts. (Figures 4.5 and 4.6)

Figure 4.5- Ferrite EC and ETD cores for power applications(Courtesy of Magnetics, Division of Spang and Co. Butler, Pa.)

4.1.6-ETD Cores

ETD cores are similar to E-C cores. They have a constant cross section for high output power per unit weight and simple snap-on clips for holding the two halves together. They also have a bobbins which provides for creepage for mains (line) isolation and have enough space for many terminals. Zenger(1984) suggests that the constant cross section of the ETD is an important attribute for high frequency and high drive levels. (Figure 4.5) These cores are available only in the large sizes and thus are not used with surface mount bobbins.

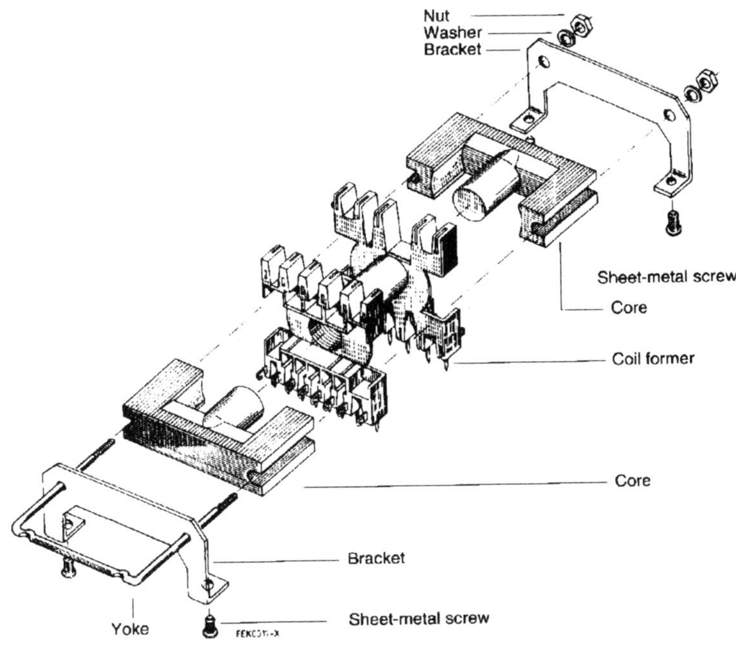

Figure 4.6- Ferrite EC cores for power applications (Courtesy of Siemens-Matsushita, Data Book, Ferrites and Accessories, 1997)

4.1.7-E-R Cores

These cores combine high inductance and low overall height. They have a round center post and surface mount bobbins available with the smaller sizes.

4.1.8-EP Cores

EP cores are a modification of a pot core but the overall shape is rectangular. A large mating surface allows better grinding and lapping, preserving more of the material's permeability. The EP core is usually mounted on its side with the bobbin below it facilitating printed circuit mounting. The best advantage of this core is in high permeability material. Shielding is very good. Some sizes of E-P cores (EP7 and EP13) are available with surface-mount bobbins with gull-wing terminals. Figure 4.7 shows an assortment of EP cores with the mounting accessories.

Figure 4.7-EP Cores for Power Application

4.1.9-PQ Cores

TDK says it stands for Power and Quality. These are one of the newest types of cores for power ferrites for switched mode power supplies. The lowest core losses in a transformer usually exist when the core losses equal the winding losses. The geometry in a PQ core is such as to best accomplish this requirement in a minimum volume. The clamp is also designed for a more efficient assembly. A more uniform cross sectional area is also achieved so that the flux density is uniform throughout the core so that the temperature will not vary much. See Figure 4.8.

4.1.10-Toroids

Toroids are sometimes used as power shapes because they take full advantage of the material permeability. Since there is no gap, leakage is very low. The toroid's main disadvantage is the high cost of winding as compared to an E or pot core. (Figure 4.9). Engelman (1989) constructed a multi-toroid power transformer that provides digital control. 40 toroids were used.

Figure 4.8- Ferrite PQ cores for power applications (Courtesy of Magnetics, Division of Spang and Co. Butler, Pa.)

Bates (1992) reported on a new SMP core technology combining new high frequency ferrite power materials as toroids in a matrix transformer that can deliver 2000 watts at 5V D.C. It has the advantage of being low profile, has low leakage inductance excellent winding isolation and higher thermal dissipation due to increased surface area.

4.1.11-EFD Cores

Probably the newest design in miniature power shapes is the EFD cores which stands for E- core with flat design. (See Figure 4.10) The center leg was flattened for the extra low profile needed for PC board mounting. Simple clips are available. As expected surface mount bobbins are available. Mulder (1990) has written an extensive application note on Design of Low-Profile High Frequency Transformers. He finds an empirical relation between effective volume and the thermal resistance of a magnetic device with which a

Figure 4.9- Ferrite toroids (Courtesy of Magnetics, Div. of Spang and Co., Butler,Pa.)

CAD program can be constructed to develop the optimum range of EFD cores for the frequency band 100KHz to 1 MHz.

4.2-EFFECTIVE CORE PROPERTIES-POWER CORE SHAPES

In Chapter 3, the inherent properties of the various power magnetic materials were listed. In most every case, the measurements reported were made on toroids or ungapped shapes. We know that most power shapes are gapped. Even the so-called "ungapped" cores have a gap in the mating surfaces. For gapped cores, the dimensions and magnetic properties must be modified to the effective parameters

The toroid was described as a closed magnetic circuit with uniform cross section. Even in a toroid, however, the magnetic path length varies from the circumference formed by the ID and to that formed by the OD. The mean length is often taken as the circumference of the average diameter [$le = \pi(d_o + d_i)/2$]. Where there is a large variation between the OD and ID, the average

Figure 4.10- EFD Cores for Power Applications

value is invalid and a more complex method involving integration of all the paths is necessary. The situation on other shaped components is usually not as simple. First, the circuit may have an air gap (intentional or that formed by mating surfaces). The permeability of the magnetic circuit will be;

$$\mu_e = \mu_o/\{1+ \mu_o l_g/l_m\} \qquad [4.1]$$

where; μ_e = Effective permeability of gapped structure
μ_o = permeability of the ungapped structure
l_g = length of gap
l_m = length of magnetic path

It is very important for us to appreciate the impact of this relationship especially in high permeability materials. For example, let us take the case of an EP core of 10,000 permeability material with no intentional gap. A separation of only 1 micron(or .00004 inches) will reduce the effective permeability to about 6700.

The effective permeability, μ_e is actually the permeability of an equivalent ungapped structure having the same inductance and same dimensions.
If there is a varying cross section of the component (such as a pot-core), then special methods are available for determining the effective length, le, the effective cross section, A_e, and effective volume, V_e, of these shapes by combining the contributions of each varying section.

4.2.1-Measurement of Effective Permeability

The effective permeability can be measured by several different methods. Impedance Bridges which separate the inductive and resistive components of an impedance are generally used for ferrites. The effective permeability is given by;

$$\mu_e = L_s l_e / .4\pi N^2 A_e \qquad [4.2]$$

4.2.2-Inductance Factor, A_L

One characterization of the inductance of a component is the inductance factor A_L. It is defined as the inductance of the core in henries per turn or millihenries per 1000 turns. This factor for a specified core can be used to calculate the inductance for any other number of turns, but we must remember that since L varies as N^2, so does A_L.

$$L_N = A_L N^2/(1000)^2 \text{ for L in millihenries} \quad [4.3]$$
$$L_N = A_L N^2 \quad \text{for L in henries} \quad [4.4]$$

The standard A_L values for pot cores are chosen from the International Standards Organization R5 series of preferred numbers. In this system, the antilogs of .2, .4 .6 .8, 1.0 and their multiples of 10 are selected numbers. Thus common A_L's are 16, 25, 40, 63, 100,160,250, 400 and so on. Some companies include A_L's from the R10 series which include antilogs of .1, .5 and .9 and have A_L's of 125, 315 and 800.

4.3-GAPPED CORES

In low power transformer or inductor applications, gapped cores were used to control the inductance and to raise the Q of the core. Although many ferrite power cores are used in the "ungapped" state, either as an E core or pot core without any intentional gap, in some situations, the intentional gap can be quite useful even in power applications. These often occur when there is a threat of saturation that would allow the current in the coil to build up and overheat the core catastrophically. The gap can either be ground into the center post or a non-magnetic spacer can be inserted in the space between the mating surfaces. The gapped core is extremely important in design of filter inductors or choke coils. We shall discuss this application later in this chapter. The basis of the gapped core is the shearing of the hysteresis loop shown in Figure 4.11a and 4.11b where 4.11a represents the ungapped and 4.11b the gapped core. The effective permeability, μ_e, of a gapped core can be expressed in terms of the material or ungapped permeability, μ, and the relative lengths of the gap, l_g, and magnetic path length, l_m :

$$\mu_e = \mu/[1+\mu_g\, l_g\, /l_m] \qquad [4.5]$$

With a very small or zero ratio of gap length to magnetic path length, the effective permeability is essentially the material permeability. However, when the permeability is high(10,000), even a small gap may reduce the permeability considerably. For a power material with a permeability of 2,000 and a gap factor of .001, the effective permeability will drop to 1/3 of its ungapped value. When each point of the magnetization curve is examined this way, the result is the sheared curve shown in Figure 4.11. Ito(1992) reported on the design of an ideal core that can decrease the eddy current loss in a coil by the use of the fringing flux in an air gap. The design includes a tapering of the core at the air gap. The reduction in temperature rise will depend on the operating frequency, the gap length and the wire diameter.

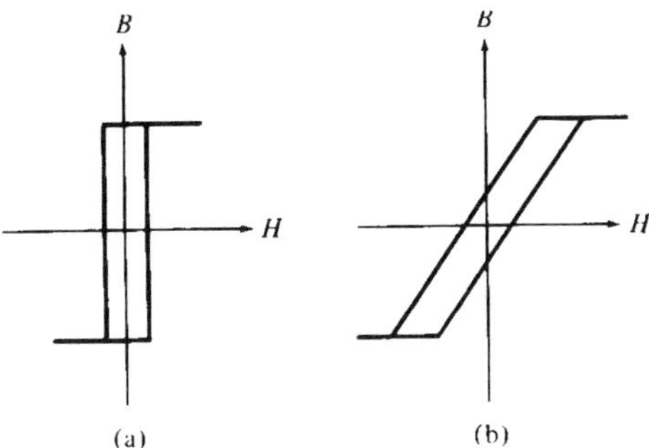

Figure 4.11-Shearing of a hysteresis loop by the application of an air gap in the magnetic circuit.

4.3.1-Prepolarized Cores

Another variation of the gapped core is one that is prepolarized with a permanent magnet. If the transformer operates in the unipolar mode and the polarity of the magnet is opposite to the direction of the initial ac drive, the starting point for this induction change will not be the remanent induction as is usually the case but a point much lower down on the hysteresis loop and in the opposite quadrant. The flux excursion will be much greater, possibly two or more times higher than the simple unipolar case. Magnetic biasing is old but the extension to this application has been described by Martin (1978) . To avoid eddy current losses, the magnet used may be a ferrite magnet often of the anisotropic variety. Shiraki (1978) reported a reverse-biased core for this purpose. Using a high-energy rare-earth cobalt magnet for the bias, he reduced the volume of the core 56% and the copper wire by a corresponding amount. Shiraki points out that the reverse-biased core has higher inductance near the normal saturation than the unbiased core. With this device, he more than doubled the volt-amp rating of the transformer. Nakamura (1982) reported a 70% increase in the figure of merit namely the LI^2. Thus, size and weight was reduced. The losses were not significantly higher under these conditions. Sibille (1982) also reported on several different geometries to implement the prepolarized core.

Prepolarized cores are especially useful in flyback and inductor applications with high DC components. Huth (1986) has described a clever way of biasing a core using orthogonal winding techniques. (See Figure 4.12).

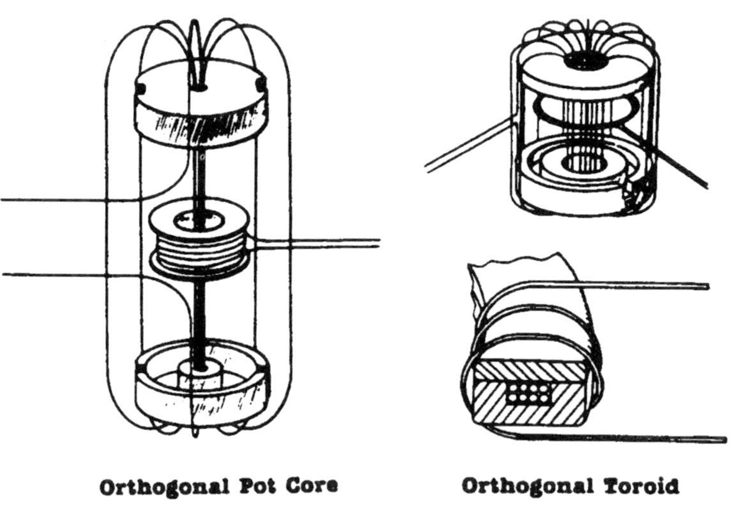

Figure 4.12-Orthogonal windings on ferrite pot cores From Huth, J.F. III, Proc. Coil Winding Conf. Sept. 30-Oct. 2, 1986

4.4-LOW-PROFILE FERRITE POWER CORES

For low power ferrite applications, the past 5-10 years have seen the introduction of low profile cores in several configurations. One reason for this change is explained in the section in which the permeability is maximizes by having the winding length large and the cross section small. This condition can be accomplished in a low profile or low height core. The other reason (also mentioned in Chapter 1) is the growing use of PC (printed circuit) boards on which to mount the magnetic cores. This method of attaching cores is even more important in the power ferrite area than in the low power telecommunications area since PC technology is increasingly placing the power supply for a circuit on the same PC board as the other circuit components. The space between the boards is one half inch so the power ferrite core must be designed to fit in that space with the bobbin and mounting hardware. The availability of low profile cores has been discussed under the sections dealing with the various core shapes. A low-profile EFD core is shown in Figure 4.13.

4.5-SURFACE-MOUNT DESIGN IN POWER FERRITES

The use of surface mount design has been used for low power ferrite applications. The motivation was the development of PC board technology surface-mount design (SMD. As with the low profile cores, the application has been widespread mostly in the power ferrite application. The use of low-profile ferrite cores can be complemented to a large degree by surface-mount technology. The two terminal mounting types used for power ferrites are the gullwing and the J-type terminals shown in Figure 4.14. The gull wing form is used when thin wire up to .18 mm in diameter is used. The J-type design is used in wire sizes greater than .8 mm. Surface mount design lends itself to high speed automatic component placement on the PC board. A surface-mount bobbin with gullwing terminals is shown in Figure 4.15. The placement on the PC board is also shown.

4.6-PLANAR TECHNOLOGY

Continuing with the low-profile design tendency particularly with PC board mounting has led to a completely new generation of cores called planar cores. Huth(1986) reported on this earlier and now, most ferrite companies offer planar cores in several varieties. Some of the arrangements are shown in Figure 4.16. Either the E-E or E-I configuration is used. The I core is actually a plate completing the magnetic circuit. In many cases the windings are fabricated using printed circuit tracks or copper stampings separated by insulating sheets or constructed from multilayer circuit boards.(See Figure 4.17)

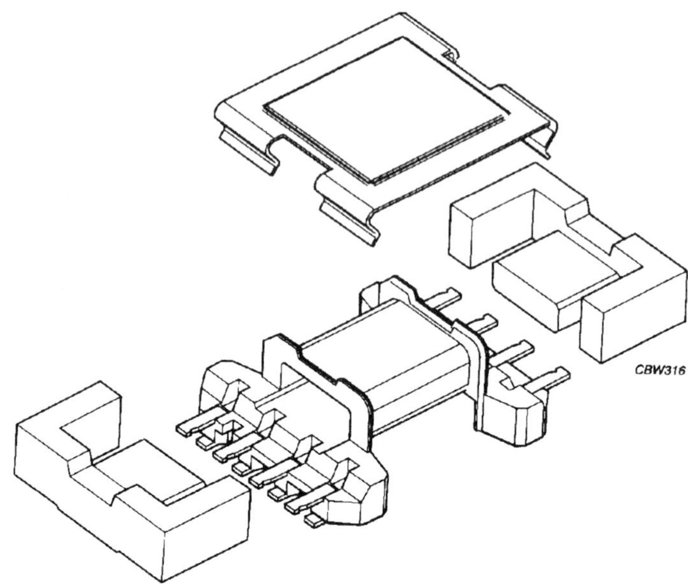

Figure 4.13-A low-profile EFD core with clip and coil former From Philips (1998)

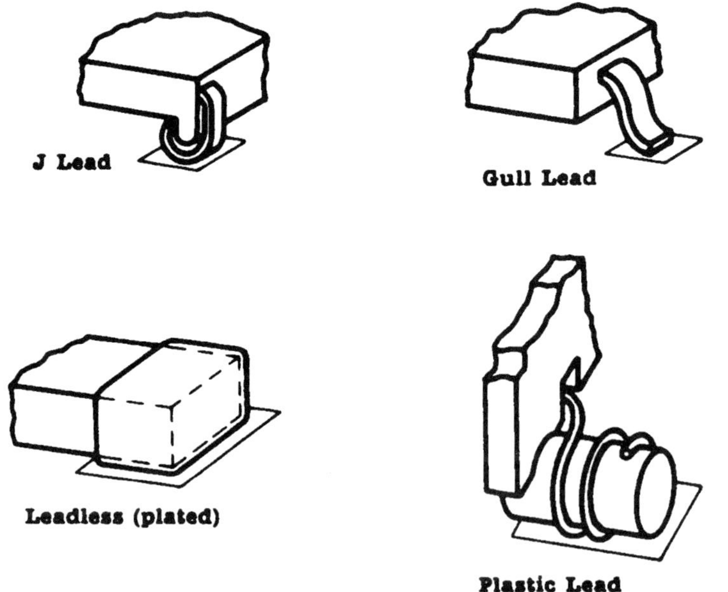

Figure 4.14-Various techniques of surface mounting of ferrite cores to printed circuit boards. From Huth, J.F III, Proc. Coil Winding Conf. Sept. 30-Oct. 3, 1986, 130

In some cases, the windings are on the PC boards with the two sections of the core sandwiching the board. Philips (1998) claims the advantages of this approach as;

1. Low profile construction
2. Low leakage inductance and inter-winding capacitance.
3. Excellent repeatability of parasitic properties.
4. Ease of construction and assembly
5. Cost effective
6. Greater reliability
7. Excellent thermal characteristics-easy to heat sink.

Yamaguchi(1992) performed a numerical analysis of power losses and inductance of planar inductors. A rectangular conductor was sandwiched with magnetic substrates. He suggested that the air gap between two magnetic substrates is an important factor governing the trade off between inductance and iron losses. Sasad (1992) examined the characteristics of planar indutors using NiZn ferrite substrates. A planar coil of meander type is embedded in one of the NiZn ferrite substrates and covered with another with a specified air gap.

A buck converter of the 10 Watt class was constructed using the inductors with an efficiency as high as 85 percent and a switching frequency of 2 MHz.Varshney (1997) has described a monolithic module integrating all of the magnetic components of a 100 Watt 1 MHz. forward converter using a plasma-spray process for deposition of the ferrite which serves as the core.

Mohandes (1994) used integrated PC boards and planar technology to improve high frequency PWM (Pulse Width Modulated Converter) performance. Estrov (1986) has described a 1 MHz resonant converter power transformer using a new spiral winding with flat cores that solved eddy current losses, leakage inductance and other problems. He also used planar magnetics and low-profile cores to cut the height and improve converter efficiency from 20 KHz. to 1 MHz. Brown (1992) replaced the traditional copper wire with a winding from the PC board or stamped copper sheet and using a low-profile ferrite core improved the performance and manufacturability of HF power supplies. Huang (1995) described design techniques for planar windings with low resistance. Three representative pattern types were explored; circular, rectangular and spiral. Gregory (1989) has described the use of flexible circuits to work with new planar magnetic structures. He claims that printed circuit inductors reduce losses and increase packing density making them an excellent choice for high-frequency magnetics. Figure 4.18 shows a collection of low-profile and planar cores.

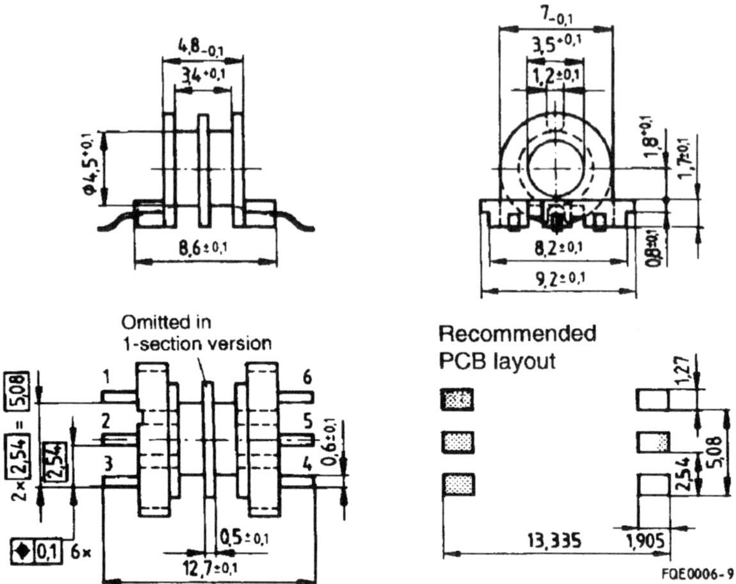

Figure 4.15- A surface-mount coil with gullwing terminals for an EP core. Also shown is the recommended PC layout. From Siemens (1997)

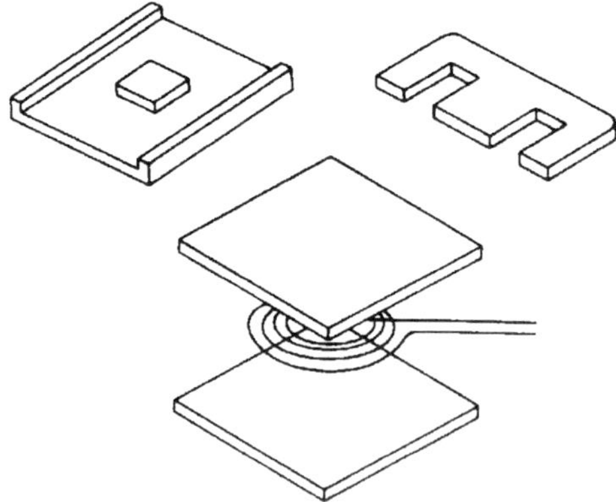

Figure 4.16-Planar ferrite cores representing a new generation of low profile cores From Huth, J.F. III, Proc. Coil Winding Conf. Sept. 30-Oct. 2, 1986

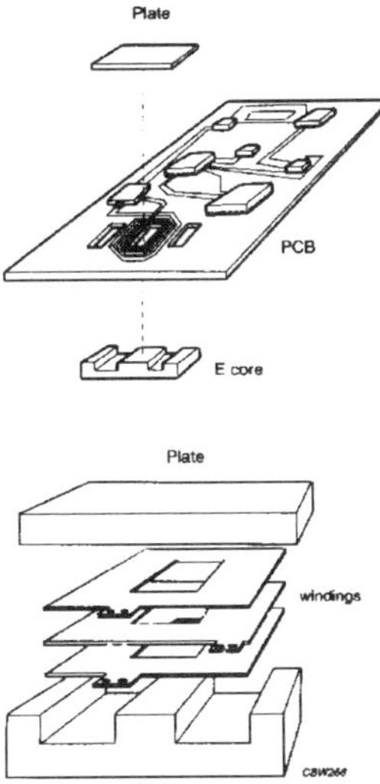

Figure 4.17 Illustrations of Planar technology for mounting cores on PC boards From Philips (1998)

Figure 4.18- A collection of low-profile and planar cores for power applications

4.7-INTEGRATED MAGNETICS

Bloom (1994) has shown the application of planar-type "integrated" magnetics wherein the transformer and inductor element can be combined on the same core with separate wing. An example of this technique is shown in Figure 4.19. The use of folded windings on printed circuit boards with flexible fold lines is shown in Figure 4.20.

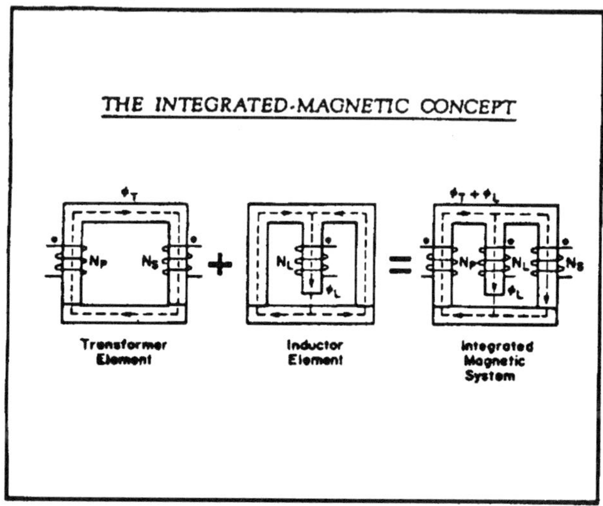

Figure 4.19-Example of Integrated Magnetic From Bloom (1994)

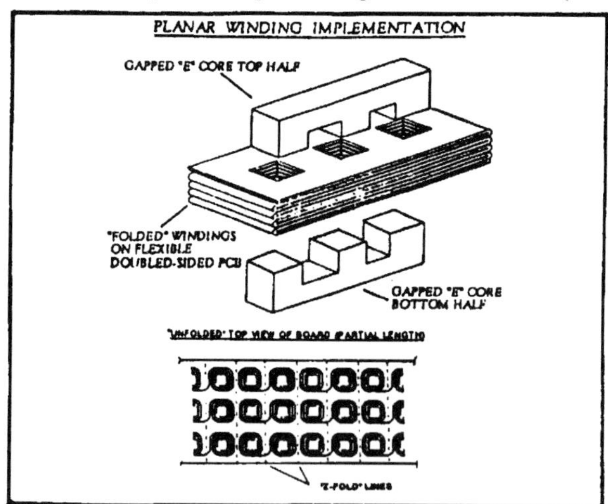

Figure 4.20- Use of 3-bobbin integrated magnetic design. The need for 3 separate bobbins is eliminated by use of folded PC board design. From Bloom(1994)

4.8- CORE SHAPES FOR METAL STRIP MATERIALS

Metal strip materials as discussed here for power electronics applications include;

1. Thin gage Si-Fe alloys (0.001-0.004 inches)
2. NiFe Alloys (Permalloys)
3. CoFe Alloys (Supermendur)
4. Amorphous Alloys
5. Nanocrystalline Alloys

The shapes of the into which these alloys are formed are;

1. Tape-Wound Cores
2. Tape-Wound Cut Cores
3. Stacked Laminations

For the power electronic applications, stacked laminations are rarely used since the metal thicknesses of laminations are normally greater than those compatible with high frequency operation. In addition, with the amorphous and nanocrystalline materials, their hardnesses make it economically unattractive to punch because of die wear. For lower frequencies and higher power, cut cores can be used successfully because of the lower winding costs of cut cores compared to tape-wound cores. That leaves the bulk of the usage of metal strip components for power electronics to tape wound cores. Unlike ferrites whose inherent magnetic properties for a single material are the same regardless of shape or size, core properties of a specific strip wound core are dependent on strip thickness. This condition arises from the lowering of high-frequency Eddy Current losses as the strip thickness is decreased. In addition, the so-called "stacking factor" or ratio of volume of metal to volume of wound core must be accounted for in flux density consideration.

4.9 CORE SHAPES OF METAL POWDER COMPONENTS

Metal powder core components of a specific base chemistry are classified according to permeability. The permeability of a particular material is determined by the physical properties including particle size, amount of insulation and pressed density. There are two main shapes of metal powder core components. They are;

1. Toroids
2. E-Cores

Although, traditionally, toroids have been the shape of choice since the low permeabilities of powder cores are additionally lowered by a gap. However, in recent years, there has been growing usage of the E-core design. Here again, the motivation has been the elimination of costly toroidal winding.

SUMMARY

This chapter has listed the shapes that are commonly used in components for power electronics. The next chapter will discuss the techniques that are used to determine the optimum size of the component. As such the material and component parameters will be integrated into the circuit requirements.

References

Bloom, G.(1989) Powertechnics April 1989, 19
Estrov, A.(1989) and Scott,I., PCIM, May 1989
Huth, J.F. III, (1986)Proc. Coil Winding Conf. Sept. 30-Oct. 2, 1986
Magnetics (1987) Ferrite Core Catalog, Magnetic Div.,Spang and Co, Butler, PA 16001
Martin, W.A.(1978), Electronic Design, April 12,1978, 94
Nakamura, A.(1982) and Ohta, J.,Proc. Powercon 9,C5, 1
Shiraki, S.F.(1978), Electronic Design, 15, 86
Sibille,R. (1981), IEEE Trans. Magnetics, Mag.22 #5,Nov. 1981, 3274
Sibille, R.(1982), and Beuzelin, P., Power Conversion International,1982, 46

Chapter 5

CORE SIZES-DESIGN CONSIDERATIONS IN POWER ELECTRONICS

INTRODUCTION

The last few chapters, the choices of components for power electronics were considered based on the circuit topology, component function, material and shape. This chapter will be concerned with the selection of the size of the core. First consideration will center on satisfying the electrical input requirements with regard to input and output voltages and currents, followed by efficiencies, regulation, temperature-rise and safety requirements. The first section will concentrate on output transformers and output inductors, followed by common-mode chokes, EMI suppression cores and magnetic amplifier components. The appendices will deal with design examples for the various functions.

5.1-DETERMINING SIZE OF THE TRANSFORMER CORE

Years ago, transformers were designed by using cut-and-try methods involving many modifications and final optimization. Such techniques are time-consuming and ineffective procedure and although some use of them remains, many design aids have been established to assist the designer in at least a close fit to the required circuit with only some minor adjustment needed. Several schemes of sizing the core and completing the circuit design are presented in this chapter. The first approach is the use of the core area-window area product that has been adopted by many manufacturers of power magnetic components. These vendors correlate the area products with certain core sizes and materials. Variations of the product area have been used by the manufacturers or authors in books. The next approach involves the use of other power specifications offered by the vendors. These may include the core losses for the cores (either per cc or gm), the core surface areas and the thermal resistances. In many cases when not all the input and output conditions are specified, some reasonable assumptions will be made in the initial designation of the core size. Since no universal scheme for sizing the core has yet emerged, the variety of different methods will be discussed.

5.1.1-Initial Considerations in Designing a Power Transformer Core

In the design of a core for a power transformer used in SMPS converters, we must take into account the input current requirements to provide the ac field to drive the core to the proper B level. This will be determined by the following equation;

$$H = .4\pi NI / l \qquad [5.1]$$

In strict operational terms, the NI of the primary winding will provide the flux variation to induce the necessary secondary voltage. This voltage is related to the operating conditions by the following equation;

$$E = 4.44\ BNAf \times 10^{-8} \qquad [5.2]$$

for sine wave with the coefficient changing to 4 for square wave. Although part of the dimensions (cross sectional area) of the magnetic core is related directly to the flux requirements imposed by the second equation, all the windings in a power core are contained inside the core. This includes the primary turns, N_p, determined by the magnetizing current equation and the secondary turns, N_s, given by the induction equation. These windings are contained either inside the window of the toroid or a U or E core or are on a bobbin surrounding the center post in a pot core. Consequently, the size of the window or bobbin winding space does directly affect the overall size of the core. Therefore, it is these two requirements that are related in the design determining the shape and size of the core.

In other words, the flux equation contains the cross sectional area of the core. The NI requirements must be met by a certain number of turns each having a certain capacity to carry a current I. Achieving a higher current may allow only a few turns with a larger cross sectional area per turn as opposed to a design carrying a larger number of turns with a smaller cross sectional area per turn. It is the product of the NI which is a measure of the total copper cross sectional area and which will determine the window area.

Therefore, there are two areas that will at first determine the size of the core. One criterion used for years by design engineers is the product of these areas that is called the Area Product, A_p, (Magnetics 1987) described by;

$$A_p = W_a A_c \qquad (cm^4) \quad [5.3]$$

Where; A_p = Area Product
W_a = Area of the window (cm^2)
A_c = Area of the window (cm^2)

Of course, the A_c is the area transverse to the flux and the W_a is the area transverse to the current flow. The area of the window is not completely usable because of the space between the wires and also the insulation thickness. Therefore, we introduce a copper-filling factor, K, which is the fraction of the window containing the copper. The total cross sectional area of the copper is given by;

$$A_{cu} = NA_w$$

where; A_{cu} = area of copper (cm^2) [5.4]

A_w = Cross sectional area of the copper wire, cm^2.

Therefore, the copper filling factor, K, is;

$$K = A_{cu}/W_a = NA_w/W_a \qquad [5.5]$$

$$N = KW_a/A_w \qquad [5.6]$$

If we multiply by A_c, we get;

$$NA_c = W_aKA_c/A_w \qquad [5.7]$$

Now, from Equation 2.2 for a square wave;

$$NA_c = E\,x10^8/4Bf \qquad [5.8]$$

Setting the two equations equal and rearranging;

$$W_aA_c = EA_w\ x\ 10^8/4BfK \qquad [5.9]$$

The area of the wire is related to its current- carrying capacity by one of several analogous factors. Traditionally, electrical engineers have spoken of wire sizes in circular mils instead of cm^2 (possibly because the number is quite small in cm^2). A circular mil is the cross-sectional area of a wire whose diameter is 1mil or .001 inches. The area is then .7854 square mils or 5.0671 x 10^{-5} cm^2 . Therefore, to convert from a A_w or possibly a W_a in cm^2 to circular mils, divide the circular mils by this last number. If, as is a common practice, the current carrying capacity of the wire is given in terms of C in circular mils/ampere, the relevant equations are;

$$C = A_w/I \qquad \text{Circular mils/ampere} \quad [5.10]$$

Then;

$$A_w = IC \qquad [5.11]$$

The input power, P_i is

$$P_i = EI \qquad [5.12]$$

If we further define the efficiency;

$$e = P_o/P_i \qquad [5.13]$$

Where;

P_o = Output power

We can then relate the output power, P_o to the Area Product, A_p;

$$W_aA_c = P_oCx10^8/4BefK \qquad [5.14]$$

If some assumptions are made about C (800-1000 circular mils/amp), e at about 80-90%, and K (about .2-.3) we can simpify the equation. Note that the K value is only the copper-filling factor only for the primary that normally occupies about 50% of the winding space. The rest is occupied by the secondary winding. Based on these assumptions, we arrive at an equation relating P_o to operating conditions with a single constant, k_1;

$$W_aA_c = k_1P_0 \text{ x } 10^{11}/Bf \qquad [5.15]$$

If the B level is set at 2000 Gausses, families of curves relating the output power, P_0 to the Area Product, W_aA_c, can be generated as shown in Figure 5.1. The various cores having the corresponding W_aA_c values are also shown. The P_o calculated from this equation is compared to the measured values in Table 18.2. The agreement is quite good. The W_aA_c ranges can also be correlated to the temperature rise of the core in operation. The table below from Magnetics Catalog(2000) gives an approximation of the temperature rise that can be expected.

Table 5.1

Approximate Expected Temperature Rise from the Core W_aA_c Values

W_aA_c RANGE	T RANGE
<.2 x 10^{-6}	<30°
.2 to 0.9 x 10^{-6}	30°C to 60°C
.9 to 3.0 x 10^{-6}	60°C to 90°C

Table 5.2-
Calculated and Measured Output Levels for Power Ferrite Components

		CORE VOLUME CM^3	CORE WEIGHT GMS	SURFACE[1,2] AREA, A_s CM^2	CORE LOSS[3] @ 100°C Watts	WaAc CIR MILS CM^2 ($\times 10^4$)	P_0 FROM P_0/W_aA_c GRAPH	MEASURED[4] OUTPUT POWER ($\Delta T = 50°C$)[5] WATTS
	42616-UG	3.52	20	24	.35	.08	30	40
	43019-UG	6.12	34	33	.6	.14	55	92
	43622-UG	10.6	57	45	1.0	.30	110	184[6]
Pot cores	44229-UG	18.2	104	66	1.8	.73	280	260
	44020-EE	17.7	87	88	1.8	.68	260	220[7]
E cores	45528-EE	42.3	212	142	4.2	1.78	630	530
	47228-EE	52.7	264	190	5.3	2.88	1,000	760
	42206-TC	1.47	6.9	25	.15	.08	25	50
Toroids	43813-TC	10.3	52	68	1.0	.68	210	230
	44925-TC	18.8	91	136	1.9	2.3	700	680

5.1.2-Other Area Product Relationships

McLyman (1982) points out a similar relationship between the area product, A_p, to the power handling capability as well as to several other important parameters used in transformer design. For current carrying capacity, a new constant, K_j is introduced making his equation;

$$A_p = (P_t \times 10^4/K_f B_m f K_u K_j)^x \qquad [5.16]$$

where: K_f = wave form coefficient
= 4.00 for square wave
= 4.44 for sine wave
K_u = window utilization factor as previous K
K_j = current density coefficient related to copper losses
x = exponent related to geometry (for pot cores, x = 1.2)
B_m = Maximum induction in Teslas (Note the change from Gausses. (1 Tesla = 10^4 Gausses)

Both K_j and x values are listed by McLyman (1982) for different core configurations

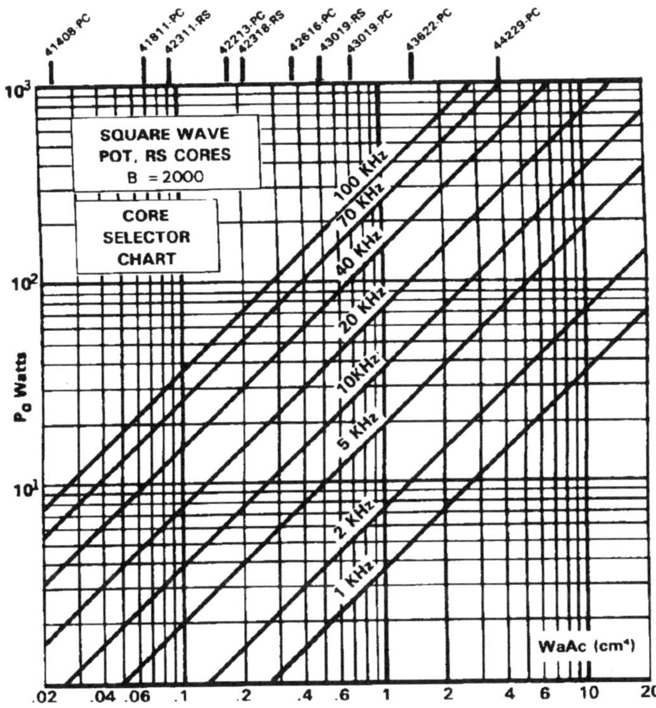

Figure 5.1- Families of curves showing output power as a function of W_aA_c at several different frequencies. At the top are listed the cores which possess the corresponding W_aA_c values.From Magnetics Catalog FC509(1997).

5.1.3-Voltage Regulation in Transformers

McLyman(1982) has also developed a new criterion and design method for transformers and inductors where the so called "regulation" is an important consideration. Smith (1985) describes regulation as "the variation in voltage from no load to full load expressed as a percentage". Thus, regulation of 5% means that 5% of the input voltage is dropped across the series resistances and reactances and the balance is transmitted to the load.

McLyman (1982) combines the power handling ability and the regulation by relating them to two constants, one a function of geometry and the other related to magnetic and electrical operating conditions. The equation is:

$$\alpha = P_t/2K_gK_e \qquad [5.17]$$

where P_t = apparent power
α = regulation in percentage

K_g = geometry coefficient
K_e = electrical coeffecient

The apparent power, P_t, is the sum of the input power, P_i, and the output power, P_o ;

$$P_t = P_i + P_o \qquad [5.18]$$

With a given efficiency, η , and in the typical D.C.-D.C Converter, the equation becomes;

$$P_t = P_o \{\sqrt{2/\eta} + \sqrt{2}\} \qquad [5.19]$$

The geometry constant, K_g is given as;

$$K_g = W_aA_c2K_u/MLT \qquad [5.20]$$

where MLT = Mean length per turn

Rather than using circular mils for the wire size, A_w and window area, W_a, these units are each given in cm^2.
The constant K_e or the electrical constant is given by;

$$K_e = 0.145\ K_f^2 f^2 B_m^2 x10^{-4} \qquad [5.21]$$

(B is in Teslas or Wb/m^2)

The current density, J, in A/cm^2 is given by;

$$J = P_t\ x\ 10^4/K_fK_ufB_mA_p \qquad [5.22]$$

McLyman gives an example of the K_g approach in the design of a transformer for a single-ended forward converter. It is shown in Appendix 2. This approach may appear long and and requires data which may not be in the manufacturers catalog. Fortunately, McClyman has supplied a compilation of the needed data in his books(1982,1988) .

Grossner (1983) has also called attention to the dependence of P_o on the area product. He uses the same approach as previously discussed with several geometrical coefficients to approximate the output power, P_o. Grossner is more concerned with the temperature rise in the calculation that is expressed as;

$$P_0 = C_2 f B g_6 (h\theta)^{1/2} \qquad [5.23]$$

where;

C_2 = a constant involving core and winding fractions and wire resistance

h = Coefficient of heat transfer
θ = Temperature rise
g_6 = Geometrical constant involving surface area, magnetic path length and wire length per turn

Thus, Grossner concludes that the power level is more responsive to increases in frequency and flux density than to an increase in the temperature rise. In practice, f is defined by the circuit and B is limited by the core material. With B and f fixed, keeping a small size and a high power level are aided by operating at the highest possible temperature rise. Because circuits may be designed to optimize different requirements, Grossner develops the parameters, g_1 - g_8, which in some combination will lead to optimization of power, inductance, and optimum power.

Smith (1985) uses the area product as a design criteria, but reduces it to Normalized core dimensions so that any size core can be calculated.

Another author using the W_aA_c approach is Pressman (1977). Here the tables of the supplier are used to approximate the core for the power level required.

DeMaw (1981) approximates the temperature rise in a core as;

$$T_{rise} = 50P_t/P_o \qquad [5.24]$$

Where P_o is not the output power as previously used but the power dissipation level within a specified core that will cause a 50°C temperature rise. P_t is the total power dissipated in an inductor including core loss and winding loss. The core loss data can be obtained from the manufacturing catalog while the winding losses can be estimated by formulae.

Watson (1986) also uses the Area product and McLyman's K_g method but derives equations based on two kinds of current density, rms and instantaneous. This distinction is important for flyback transformers.
Although the W_aA_c approach is merely a starting point in the design, many other factors may have to be considered in finalizing the design. With ferrites in power supplies some of these factors are as follows;

1. Max temperature of the core (<100°C.)
2. DC imbalance
3. Magnitude and linearity of magnetizing current
4. Magnitude of transient current
5. Under transient loading, need to limit B_m to avoid saturation

5.1.4-Other Transformer Design Techniques

Snelling(1988) has divided the design of power transformers into several different categories. They are:

1. Winding loss limited
2. Saturation limited
3. Regulation limited
4. Core loss limited
5. High frequency limited

Items numbered 2-5 are discussed previously. Snelling has added the other two. He discusses them each separately, stating that, in general, they come into play as the frequency of operation increases.

5.1.4.1-Winding-Loss-limited Design

This is almost the same situation we have been discussing using the W_aA_c approach or the wire current density approach. As in the previous cases, there is also some limit on the B level. However, here the only real source of dissipation is the winding loss and, depending on the size of the transformer, the treatment is only applicable to a frequency of about 5-10 KHz. The equation for the input power, P_i , Snelling(1988)gives as;

$$P_i^{\,2} = P_w f^2 B_e^{\,2}/mk_1 \qquad [5.25]$$

Where; m = Fractional increase in the resistivity of the copper over that at

25°C

and;

$$k_1 = 2\rho_c l_w/\pi^2 A_e^{\,2} A_w F_w$$

l_w= mean length of a turn of the winding
F_w= Winding factor of the copper

5.1.4.2-Regulation-Limited Design

In this category, the regulation can be a constraint on the winding loss listed above. Again, the winding loss is really the only source of dissipation, that is, the core loss, P_c is much smaller than P_w at these frequencies. The voltage regulation, , is given as

$$\alpha/100 = P_w/P_i \qquad [5.26]$$

Then; $P_i = (f^2 B_e^2 \alpha / mk_l\ ) x100$ [5.27]

Since the core loss is negligible, we can ignore it and the output power is given by;

$$P_0 = P_i - P_w\ = P_i[1- \alpha/100] \qquad [5.28]$$
$$= f^2 B_e 2(\alpha - .01\alpha\ ^2)/mk_l\ x100 \qquad [5.29]$$

B_e^2 is limited by the saturation flux density of the material. The other parameters are given so that k_l can be calculated. Snelling (1988) presents a table of the values of k_l in his book for many power core sizes in so that the choice can be made. Most ferrite transformers are not regulation limited.

5.1.4.3-Saturation-Limited-Design

As the flux density increases, the hysteresis curve will flatten out as saturation is approached. When this happens, the incremental permeability drops sharply. In this case, the impedance (or inductive reactance) becomes quite small and the current, therefore, increases. See Figure 5.2. The manufacturer will often give limit values for the maximum flux density that the designer should not exceed. Since the saturation drops at higher temperatures, the saturation value at the operating temperature should be examined in this regard. For operation at 100°C., the value of 3200 gausses would be a real maximum.

The constant k_l is still a valid constant for saturation-limited designs. Since the efficiencies at these lower frequencies are close to 100 percent, the output power, P_0, can be considered the same as the input power, P_i. Therefore;

$$P_o^2 = P_w f^2 B_e^2 / mk_1 \qquad [5.30]$$

If the flux, ϕ^2, is used instead of B_e^2, the factor k_2 is used where $k_2 = k_1 A_c^2$. If core loss is included in the saturation limited case, the equation for the square wave drive becomes;

$$P_0 = (P_t - P_c)^{1/2} k_3 F_w^{1/2} f \quad \text{Watts} \qquad [5\text{-}31]$$

where; P_t =Total Losses = $P_w + P_c$ [5-32]

P_t is also listed in Snelling's (1988)Table 9.3. At low frequencies, the core loss is less than half the total loss and so may be set to 0 especially because of the square root dependence.

$$k_3 = (A_w / m_c l_w)^{1/2}\, \phi \qquad [5.33]$$

Again, Snelling lists the values of k_2 and k_3 in Table 9.3 of his book (1988). In addition the value of P_t which is the sum of all the losse (winding and core) is also given in the same table. The permitted temperature rise in this table is 40°C. If the proposed temperature rise is different, the new value of B_e or can be recalculated from the thermal resistance, R_{th}.

At low frequencies, where core loss is a small fraction of the total loss, the output power is proportional to f. If the operating variables such as frequency, temperature rise and copper factor are assumed, the power handling values for a given core can be given. Most manufacturers provide such information. The values of P_0 are similar to the ones derived earlier from the area product technique.

5.1.4.4 -Core-Loss-Limited Design

Traditionally, design of a transformer is optimized by making the winding losses equal to the core losses. It has generally been taken as a device. Other calculations place the division as;

$$P_c = (2/n)P_w \qquad [5.34]$$

When n=2, the two losses are indeed equal. However at higher frequencies, the value of n is between 2 and 3. For core loss limited designs assuming the output power, P_0, is equal to the input power, P_i, the equation for the core loss, P_0, for a square wave is;

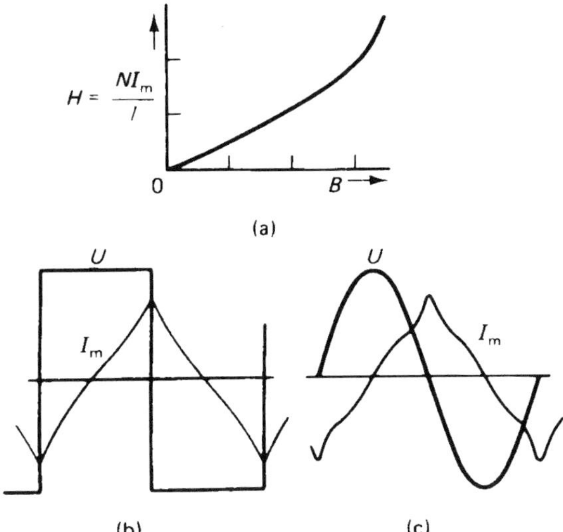

Figure 5.2-The effect of increasing induction on the field and thus the current associated with it as a ferrite core approaches saturation-(a) The increase in slope of H near saturation;(b)The voltage and current waveforms(Note the sharp increase in the slope of current at the end of the pulse. The drive is square wave. (c) The same with sine wave drive.

$$P_0 = [P_t/(1+2/n)]^{1/2} \text{ x } k_3 F_w^{1/2} f\, \phi_{p\text{-}p} \quad [5.35]$$

where K_3 is defined as before.

With P_t and k_3 given in the table, the output power can be given in the case of a core-loss limited design. To check the flux density, the manufacturers' graphs showing core loss as a function of frequency and flux density can be consulted. When P_t is known, the core loss can be estimated from the previous equation or just set to 1/2 P_t. From this core loss and the operating frequency, the B value can be read off the graph. If there is a varying cross sectional area of the core, the equation is modified as such for square wave;

$$P_o(1+2/n)^{1/2}]/2fBF_w^{1/2} = P_tA_w/m\ l_w]^{1/2} \text{ x } A_{min} = k_4 \quad [5.36]$$

The values for k_4 are also tabulated in Snellings Table 9.3.

Based on the input design specifications, k_4 and B can be calculated from the minimum area and a core can be chosen. For the division of losses, a value of 2.5 is typical. For a fully-wound transformer, F_w can be set at .5. With these assumptions, k_4 can be written as;

$$k_4 = 0.845P_0/fB \quad \text{for sine wave} \quad [5.37]$$

and

$$k_4 = 0.949P_0/fB \quad \text{for square wave} \quad [5.38]$$

The manufacturers' data is certainly a good way to check the core loss assumptions. These design methods are useful in initially picking a core and modifications must be made if one or more condition is not met.

5.1.5-Power Ferrite Design from Vendors' Catalogs

The vendors of ferrite cores have proposed several different design schemes. The one used in the Magnetics Catalog(1987) has been discussed. Design methods described by other ferrite vendors are:

5.1.5.1-Philips-(Yageo)

In the case of Philips catalog, the throughput power, P_o, information is supplied in the form of graphs of the P_o and output voltage, V_o for each type of converter. Such a graph is shown in Figure 5.3. In addition , the performance factor (f x B_{max}) is graphed as a function of frequency for their power ferrite materials. A graph of this type is given in Figure 3.11.

5.1.5.2-Epcos (Siemens)

Epcos lists the output powers for each power shape in several power materials and for each converter type. The output powers are given at a typical frequency and a cut-off frequency for each material. A portion of this table is given in Table 5.3. The material-specific values on which the table values are based were taken from the maximum temperature rise for each material given in Table 5.4 and the thermal resistance for each size and shape of core that is listed in Table 5.5.
The total core losses are related to these factors by ;

$$P_{V,\,tot} = \Delta T / R_{th} \quad [5.39]$$

The values of output power are obtained from the following formula;

$$P_{trans} = C.\frac{PF}{\sqrt{P_V}}.\frac{\Delta T}{R_{th}}.\sqrt{\frac{f_{Cu}}{\rho_{Cu}}}.\sqrt{\frac{A_N.A_e}{l_N l_e}}$$

Where PF = Performance factor
P_v = Specific Core Loss

ΔT = Temperature Rise °K
R_{th} = Thermal Resistance
f_{Cu} = Winding Factor
ρ_{Cu} = Wire Resistance.
A_N = Winding Cross-Sectional Area
A_e = Core Cross-Sectional Area
l_N = Winding length
l_e = Effective Magnetic path length

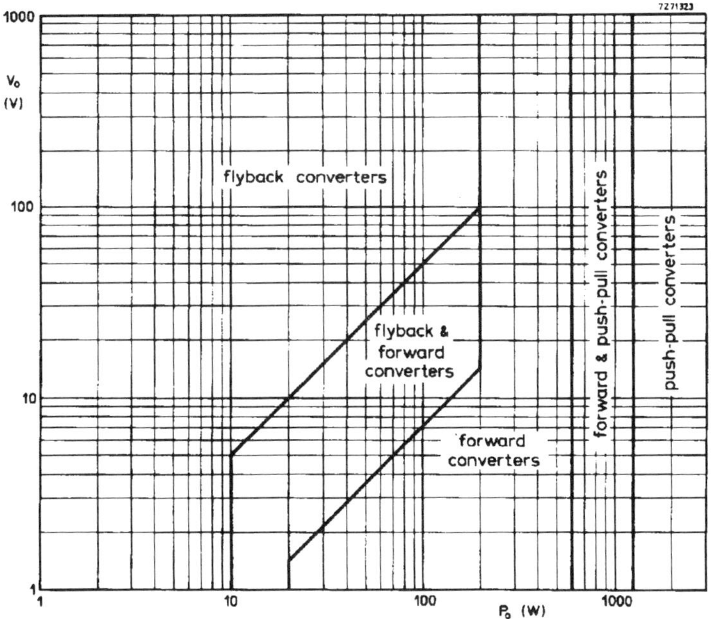

Figure 5.3-Suggested out power, P_o and output voltage for the different types of converters. From Philips (1998)

The assumption is made that the temperature rise and the losses in the core are evenly distributed. The application area for flyback transformers were restricted to 150 KHz. . The overtemperature, ΔT is the sum of the temperature rises resulting from the core and winding losses. The maximum flux densities were <200mT for flyback converters and <400mT for push-pull converters.

Table 5.3-Power handling capabilities of various shape cores & materials

					Power capacities					
					Push-pull converter		Single-ended converter		Flyback converter	
					C = 1 [1]		C = 0,71 [1]		C = 0,62 [1]	
Core shapes	Material	Version (LP = Low profile)	f_{typ} kHz	f_{cutoff} kHz	P_{trans} (f_{typ}) W	P_{trans} (f_{cutoff}) W	P_{trans} (f_{typ}) W	P_{trans} (f_{cutoff}) W	P_{trans} (f_{typ}) W	P_{trans} (f_{cutoff}) W
EFD25/13/9	N59	Normal	750	1500	311	417	221	296	193	258
	N49		500	1000	196	263	139	187	122	163
	N67		100	300	175	280	124	199	109	173
	N87		100	500	242	482	172	342	150	299
EFD30/15/9	N59	Normal	750	1500	401	343	285	244	249	213
	N49		500	1000	253	544	180	386	157	337
	N67		100	300	226	365	160	259	140	227
	N87		100	500	312	630	221	447	193	390
U cores										
U15/11/6	N27	Normal	25	150	31	81	22	58	20	50
U17/12/7	N27	Normal	25	150	37	97	26	69	23	60
U20/16/7	N27	Normal	25	150	74	161	52	114	46	100
U25/20/13	N27	Normal	25	150	198	432	141	306	123	268
UU93/152/30	N27	Normal	25	150	2527	5508	1794	3910	1567	3415

1) Numerical data are stated in accordance with the publication "Effect of the magnetic material on the shape and dimensions of transformers and chokes in switched-mode power supplies", G. Roespel, Siemens AG München, J. of Magn. and Magn. Materials 9 (1978) 145-49

From Siemens (1998)

Table 5.4- Maximum Temperature Rise and Typical and Cut-off frequencies for various ferrite power materials

	ΔT_{max} K	f_{typ} kHz	f_{cutoff} kHz
N59	30	750	1500
N49	20	500	1000
N62	40	25	150
N27	30	25	100
N67	40	100	300
N87	50	100	500
N72	40	25	150
N41	30	25	100

Table 5.5-Thermal Resistances for Main Power Transformer Shapes

Core shapes	R_{th} (K/W)	Core shapes	R_{th} (K/W)	Core shapes	R_{th} (K/W)
E 20/6	50	ETD 29	28	PM 50/39	15
E 25	40	ETD 34	20	PM 62/49	12
E 30/7	23	ETD 39	16	PM 74/59	9,5
E 32	22	ETD 44	11	PM 87/70	8
E 40	20	ETD 49	8	PM 114/93	6
E 42/15	19	ETD 54	6	U 11	46
E 42/20	15	ETD 59	4	U 15	35
E 47	13	ER 42	12	U 17	30
E 55/21	11	ER 49	9	U 20	24
E 55/25	8	ER 54	11	U 21	22
E 65/27	6	RM 4	120	U 25	15
EC 35	18	RM 5	100	U 26	13
EC 41	15	RM 6	80	U 30	4
EC 52	11	RM 7	68	U 93/20	1,7
EC 70	7	RM 8	57	U 93/30	1,2
EFD 10	120	RM 10	40	UI 93	5
EFD 15	75	RM 12	25	UU 93	4
EFD 20	45	RM 14	18		
EFD 25	30				
EFD 30	25				

5.1.5.3-AVX- (Thomson)

Thomson(1988) presents charts of average wattage for the various size and shape power cores listed according to inverter type. The frequencies are 25KHz(2000 Gausses), 100 KHz(1000 and 1200 Gausses).

5.1.5.4-TDK

TDK lists the calculated output power under the specifications of each type of power core. These power levels are given at 50 and 100 KHz for their standard power materials. These power levels are given for the forward converter mode. TDK also gives the power losses for each power core. The conditions are 25 KHz(2000 Gausses) and 100 KHz(2000 Gausses). In an older catalog, the temperature rise was also plotted against the power loss for each core. See Figure 5.4.

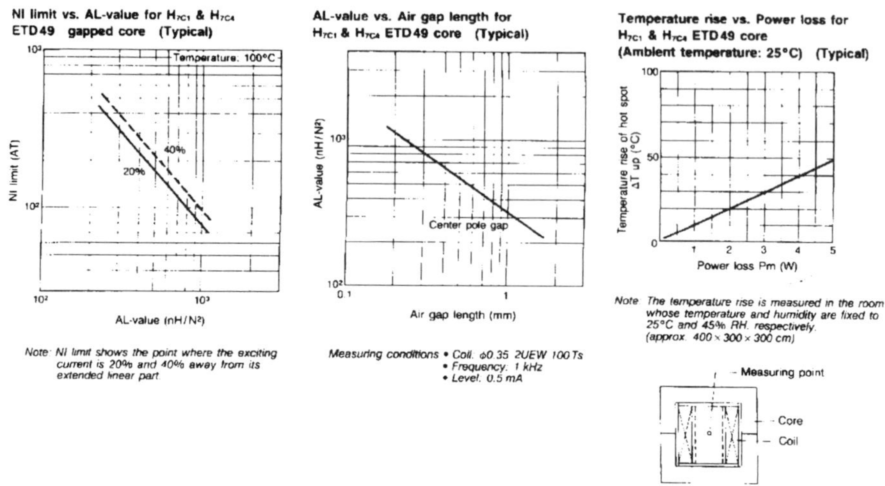

Figure 5.4-Listing of calculated output power and power loss for several different ETD cores. The temperature rise associated with that power loss is also listed. From TDK Ferrite Cores Catalog BLE873-001C(1987)

5.1.6-Ferrite Component Design-Power Transformers

The practical design given by ferrite vendors of the power transformer may vary considerably from company to company and from engineer to engineer. Usually, previous experience in the field will provide a good starting point. In addition, vendors often supply appropriate tables, graphs and figures to assist in the design. In this chapter, we will discuss some of the practical methods suggested. Recent safety requirements entering into the design will also be covered. As is the case in many higher-frequency digital devices, consideration must be given to the electromagnetic interference, both on the transmission and reception sides. Filters are therefore necessary additions to the design. Incidentally, these also use ferrite components.

5.1.6.1-Creepage Allowance

The design of a transformer that must satisfy the safety requirements of line or mains isolation provides for an 8 mm. gap between the primary and secondary winding. This is most often accomplished by leaving a space of 4mm on each end of the bobbin winding space of both the primary and secondary windings. This type of construction is shown in Figure 5.5 and 5.6 taken from Snelling(1988). Many of the core selection curves including some of Snelling's (1988) take this into account in the transformer

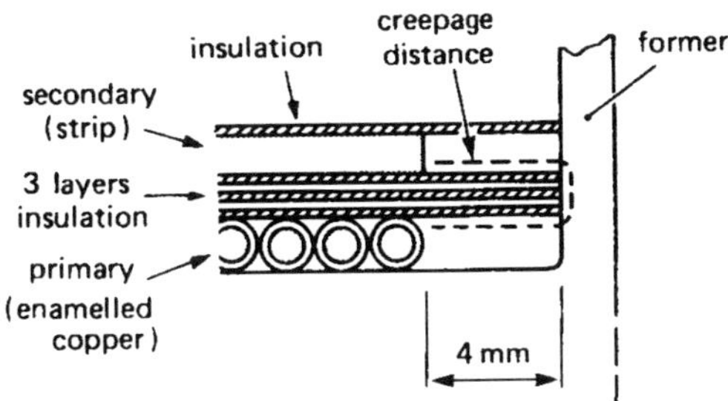

Figure 5.5-The arrangement of a winding and insulation to provide the required 8mm. "creepage" distance. From Snelling(1988)

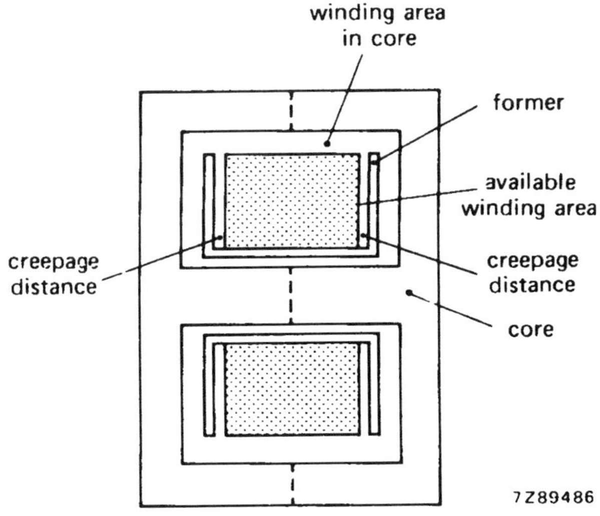

Figure 5.6.- Cross section of a ferrite core and winding showing the provision for "creepage" distance. From Snelling(1988).

design. The specification is either spelled out in the German VDE 0806 or the International IEC 435 documents. The increase in output power, P_0 without this allowance is given by;

$$P_{on} = P_{oc}\ [\text{Full } W_w/(\text{Full } W_w - 8\text{ mm.})] \qquad [5.40]$$

Where; W_w= Winding width (mm.)

and c = core with creepage allowance

n = core with no creepage allowance

This amounts to about 25% increase for small cores and about 10% for large cores. Therefore, this is an important design consideration, especially if the line or mains voltage in the country of use is 230 V. or more.

5.1.6.2- Effect of Core Size-Dimensional Resonance

For large high power reactor and transformer cores, TDK has, in addition to the previously cited PC40, another material PC22 with a slightly higher saturation value. The high power cores are available in T (Toroids), UU, EC, EIC, PQ, E, EI, PT and SP cores.

For large cores, TDK points out that the designer has to be aware of dimensional resonance. With operating frequencies rising to the megahertz region, this effect is becoming more important. The formula for the wavelength at which dimensional resonance occurs is given as;

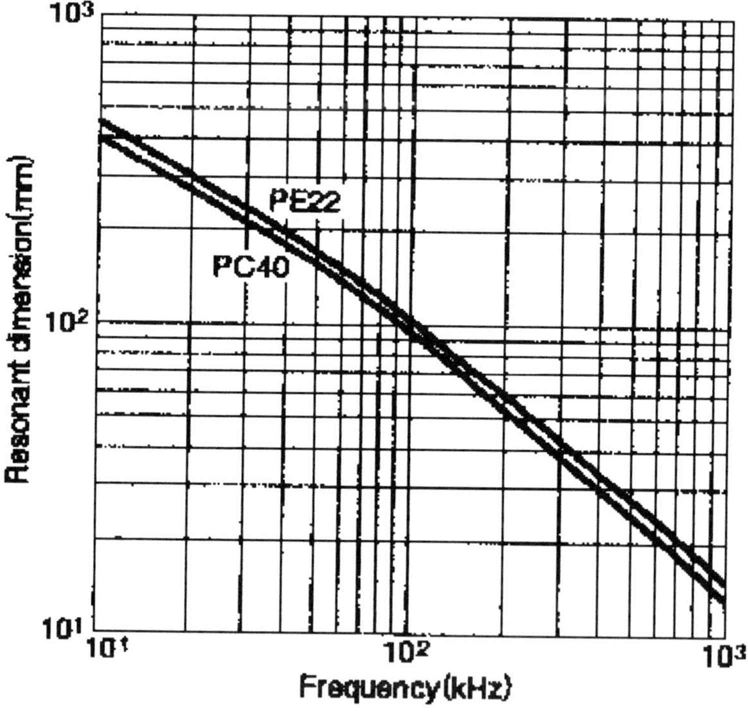

Figure 5.7 -Dimensional Resonance in mm. As a function of frequency

$$\lambda = c/\sqrt{\mu_r e_r}$$

where c = Speed of light = 3×10^{10} cm/sec
λ = wavelength in cm.
μ_r = relative magnetic permeability
ε_r = relative permittivity

Figure 5.7 shows the resonant dimensions (mm) as a function of frequency. As the effective permeability reduces sharply at dimensional resonance, the operational frequency should be set to avoid this effect. As μ_e can be reduced by the insertion of an air gap, high frequency usage can be used despite the effect.

5.2.-THERMAL CHARACTERIZATION POWER FERRITE CORES

Since the temperature rise, θ, is an important factor in the operation of the transformer in a switching power supply, it is useful to have a figure which relates the temperature rise to the power loss in a wound component. This parameter is known as the thermal resistance, R_{th} (Bracke 1982) and is given by;

$$R_{th} = \theta / P_1 \qquad [5.41]$$

Where; θ = Temperature Rise in °C.
P_o =Power loss in a wound component, Watts

The R_{th} is a function of the material as well as the shape and size of the core, the mounting method and the insulation used. It does not depend on the ratio of winding to core losses and so can be measured by noting the temperature rise as a function of the DC power loss dissipated in a winding and core . This figure can then be used at high frequency operation as done by Bracke (1982). He measured the temperature rise at D.C. and also at 50 and 100 KHz. The thermal resistance was the same at the varying frequencies. An inverse relationship between R_{th} and core size was also shown. As expected, the cores in which creepage allowance was included in the design had larger R_{th} than the same cores without the creepage allowance.

Hess (1985) found a correlation between the thermal resistance and the thermal conductivity. He found that the special power materials with the presence of CaO at the grain boundaries increased the thermal conductivity and reduced the thermal resistance. Heavier cores here also had lower thermal resistance and the ETD was particularly good.

Snelling (1988) suggests a safe value for heat loss from the surface of the core to be 300μW/mm^2 or .2W/in^2. He, therefore, recommends the cores to be broad and thin.

5.3-WINDING LOSSES

Although copper losses are not a magnetic phenomenon, the windings that produce them also carry the currents that create the fields in the primary and sense the induced voltage in the secondary. They must therefore be considered in the design. The number of turns is discussed under core losses and completion of the design.

Smith (1983) has written an algorithm for the optimum copper losses, the core losses and the sum of the losses which provides the dimensions of the cross sectional area and window area. The core loss is estimated from the equation:

$$P/V = C_1 B^{2.5} \quad \text{(loss/volume)} \qquad [5.42]$$

Both the simplification by removing the frequency dependence and concentration on toroidal shapes makes the procedure more specific to tape wound cores than ferrites. Nevertheless, the approach can be used for ferrites.

More to the point of ferrites is an article by Carsten (1986) which contains a very complete discussion on high frequency conductors in switched-mode power supplies. This article fully discusses the origin of skin, proximity and related effects. The high frequency design is especially applicable for forward and flyback converters where the inputs are below 50V. Carsten also gives a method of calculating the losses in a particular winding. The article is aimed at pulse currents with more attention than duty-cycle modulated square waves. Other wave-forms can be calculated if the harmonic content is known.

Snelling's(1988) book also has a very complete section on the design of windings.

5.4-COMPLETING TRANSFORMER DESIGN - WINDING DATA

Once the appropriate core is picked based on the limitations of saturation or core loss, the number of turns and wire size is chosen for the primary and secondary windings;

$$N_p = E_p \times 10^8/4BNA_c f \qquad [5.43]$$

$$N_s = E_s N_p/E_p \qquad [5.44]$$

and $$I_p = P_{in}/E_{in} = P_{out}/eE_{in} \qquad [5.45]$$

$$I_s = P_o/E_{out} \qquad [5.46]$$

The wire size can be obtained by referring to the winding area of the bobbin, the number of turns, the copper fill factor and a table of the wire sizes.

$$F_w W_a = N_p A_{wp} + N_s A_{ws} \qquad [5.47]$$

For a first approximation, the available window area is assumed to be divided equally between the primary and the secondary so that each winding size can be evaluated separately. The copper fill factor can be assumed to be .4 for toroids and .6 for pot cores and E cores. The number of turns fitting in the winding space of individual cores can by found from the winding tables published by the vendor and the window area published under each core. From the wire tables, the resistance per unit length of winding can be found for a particular wire size. This unit resistance, when multiplied by the length of winding (number of turns times the average length per turn given in the core data), yields the total resistance of each winding. Then the winding loss per winding can be calculated by;

$$P_{wp} = I_p^2 R_p \qquad [5.48]$$
$$P_{ws} = I_s^2 R_s \qquad [5.49]$$

The total winding loss,

$$P_w = P_{wp} + P_{ws} \qquad [5.50]$$

This can be compared with the optimum ratio of winding loss to total loss;

$$P_w = P_t/[1+2/n] \qquad [5.51]$$

If the calculated loss is close to the optimum, the design is probably pretty good. Otherwise further modifications in the winding or core size may be necessary.

5.5-VERY-HIGH-FREQUENCY POWER FERRITE OPERATION

In the last five years, the frequency of switching power supplies has increased dramatically. While initially it was 25 KHz., the present state of the art is 100 KHz., and new designs for the 200-500 KHz. range extending upwards to 1 MHz. are being developed and as a result, there has been a large reduction in the size and weight of the transformers. Engineers working with ferrite materials have responded with materials capable of operating at these frequencies. This has been done as described in a previous chapter by a combination of chemistry and microstructural improvements. We may be approaching the limit of operation of MnZn ferrites and new chemistries will be forthcoming in the NiZn materials. Snelling(1989) points out that problems with the conventional switching supply using pulse width modulation techniques is also reaching its limit because of circuit problems such as switching transients and the radio interference caused by harmonics. Snelling (1989) predicts a limit of 500 KHz. due to these effects. He also predicts that resonant power conversion may take over at the higher frequencies.

For materials up to this limit, he advises the minimization of m in the equation

$$P_m = kf^m B^n \qquad [5.52]$$

This would include materials of lower permeabilities. There should be an optimum composition for minimum power loss at a given frequency.
As previously mentioned, the design of ferrites for very high frequency application should also involve lowering of the B_{max} of the material because the flux density dependence (n) is of higher order than the frequency dependence. Therefore for the same power level, decreasing the losses by using lower flux density has more leverage than increasing the losses by the higher frequency. The ratio of eddy Current to magnetic losses is shown in Figure 5.8. Even at 250 KHz, the ratio of Eddy Current to magnetic losses, while increasing is still only about 20%. Another design factor in the higher frequencies is the advisability of operating at as high a temperature as is feasible. Better design of cores such as those of the planar type is also being used as this increases surface area for removal of heat from the ferrite.

In a recent paper, Buethker(1986) has considered the breakdown of losses in power ferrites. He lists them as

Hysteresis Losses, $P_{hyst} = C \times f^x B^y$ [5.53]
Eddy Current Losses, $P_{e.c.} = .8f^2B^2A_e/\rho$ [5.54]
Residual Losses, $P_{res} = 2.5 \times 10^{-3} f B^2 \tan \delta / u$ [5.55]

At 100 KHz., the hysteresis loss is predominant (Figure 5.9) and a large reduction in these losses in 3C85 over 3C8 accounts for a significant overall loss reduction. However, at 400 KHz., the 3C85 (Figure 5.10) shows a greater increase in eddy current and residual losses which now can be lowered by rather drastic changes in microstructure in the new material 3F3 (Figure 5.12b) so that now again, the hysteresis losses are predominant. Historically, it seems that when the state of the art of power supply design requires power ferrite material for higher frequencies, the ferrite designer produces materials with lower eddy current losses to meet the challenge.
Earlier, we discussed the performance factor PF_{200} proposed by Stijntjes(1989) which is the product ($B_m f$) of the frequency, f, and the B level which in a

given material will give losses of 200 mW/cm^3. Figure 5.11 shows the PF_{200} of 4 different materials as a function of frequency. The optimum operating

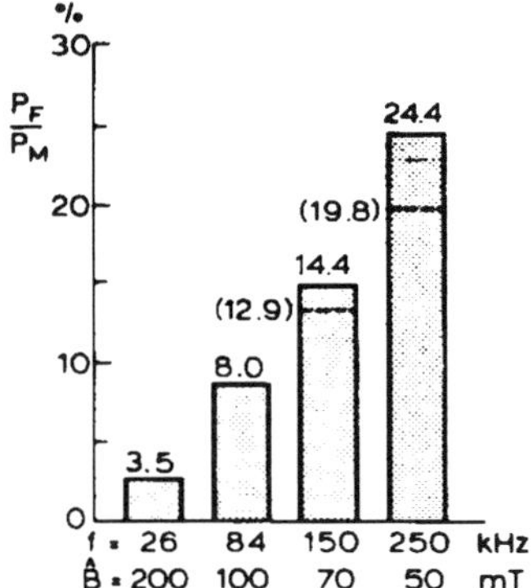

Figure 5.8-Ratio of Eddy current to magnetic loss in an ETD core. From Snelling(1989)

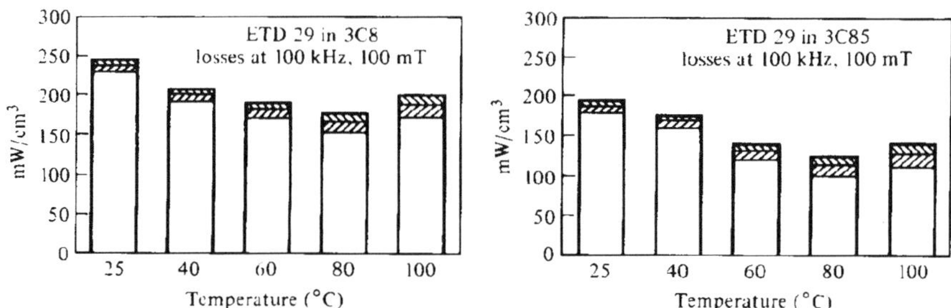

Figure 5.9-Breakdown of core losses into Eddy current, Hysteresis and Residual Losses at 100 KHz.for two ferrite materials. From Buthker,C.,and Harper,D.J., Transactions HFPC, May,1988,p.186.

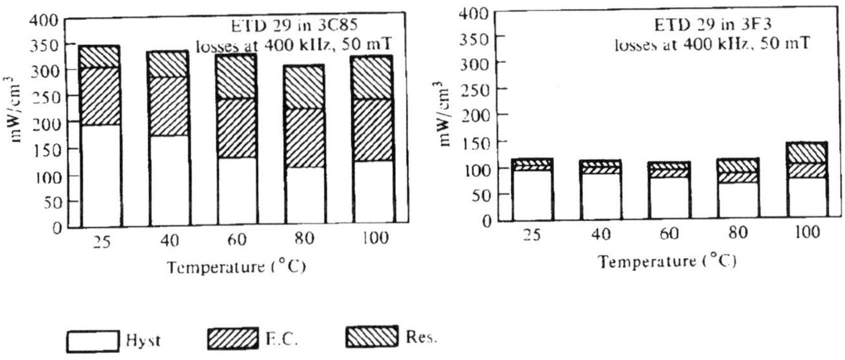

Figure 5.10- Breakdown of core losses similar to Figure 18.54(a) but at 400 KHz. From Buthker(1988)

for a material with regard to power occurs where the curve is a maximum. For material A(MnZn Ferrite with Ti and Co additions), the maximum PF_{200} is 35,000 at .5 MHz(500 KHz). The corresponding figure for material C(NiZn+Co[fine]) is 110,000 at 30 MHz. It would appear that operation at 30 MHz with material C would be more desirable but other considerations such as availability of semiconductors and the cost of the NiZn ferrite appear to be more important. Thus, for the present, improved MnZn ferrites are the major power materials.

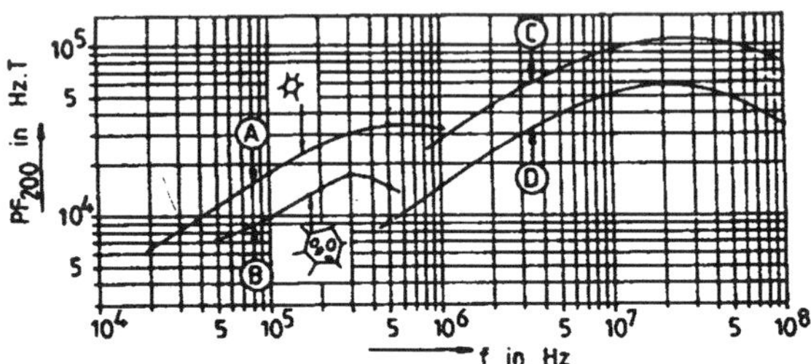

Figure 5.11-The Performance Factor, PF_{200}, for several types of MnZn and NiZn ferrite materials as a function of frequency. From Stijntjes (1989)

Recent papers on high frequency transformers or materials include; Sano(1988),Finger(1986),Kamada(1985),Carlisle(1985),Hiramatsu(1983),Martin(1984), Cattermole(1984), Shiraki(1890), Kepco(1986), Mochizuki (1985),Schlotterbeck), and Zenger(1984).These references may be found at the end of this chapter in Appendix 5.6

5.6-FERRORESONANT TRANSFORMERS

We have spoken of resonance as it relates to low level linear ferrite components. Here a series or parallel combination of an inductor and capacitor acted as an LC circuit for frequency control in low level filters. The term resonance (more properly, ferroresonance) here has more of a connotation of resistance to changes in the input voltage and current by storing energy in the resonant circuit. As a matter of fact the first uses of ferroresonance was in the construction of a constant-voltage 60 Hz. transformer by Sola. In power supplies, an important use of the ferroresonant transformer is as a regulator. The early 60Hz transformers have given rise to the high-frequency type that, as noted earlier, may be even more useful at the highest frequencies than the conventional switching transformer design.

As a high-frequency power inductor, the ferroresonant transformer has a quite different function . For one thing, the magnetic circuit is non-linear and because of the high currents and fields, operation is close to saturation. Most often when used as a power inductor, it is necessary to insert an air gap or spacer to avoid saturation.

Figure 1.19 shows a simple ferroresonant regulator that consists of a linear inductor, L_1, a non-linear saturating inductor, L_2 and a capacitor, C_1 in parallel with L_2. It is the latter two that form the ferroresonant circuit that controls the input voltage. The input energy is stored in L_1 and the resonant circuit acts to pass a uniform voltage to the load. Although the linear transformer may be of the typical power ferrite found in transformers, the saturating transformer is quite different. In addition to the usual attributes of power ferrites, it should possess a rather square hysteresis loop. The squareness ratio, $B_r/_{Bs}$ should be over 85%. The permeability over the linear portion of the loop should be as high as possible with the saturation permeability quit low ($\mu = 20\text{-}30$)

McLyman(1969)has shown how a high frequency ferroresonant transformer, tuned to about 20 KHz. can be used to stabilize high frequency inverters.

5.7-DESIGN OF POWER INDUCTORS

Some of the same factors of concern in transformers are applicable in inductors. The main difference is that the main power component in the power inductor is the DC current with the ac component only about 10% of the DC component. The DC bias displaces the point of operation to a point closer to saturation. There must be sufficient inductance remaining to effect the filtering of the ac present. The amount of power that can be handled and thus, the temperature rise are important. Regulation needs may also be present. To specify the inductor, we must have a minimum inductance and a maximum current. The stored energy of the inductor is given as;

$$E_i = 1/2\ LI^2 \qquad [5.56]$$

In the design of an inductor, we must specify the minimum inductance, L_{min} and the maximum current, I_m. This permits us to calculate the LI^2 product. The conventional manner and time tested design technique is through the use of Hanna Curves (Hanna 1927). An example of a Hanna curve is given in Figures 5.12 or 5.13 for a specific material. Here, LI^2/Volume is plotted against H. Also shown is the ratio of the gap that must be inserted in the core to the magnetic path length. Although the volume is not known at the start of the design, a trial number can be used and then by iteration of the calculation, the optimum core chosen. A point on the center of the scale can be a starting point. From the chosen core, the core volume is used to recalculate LI^2/V and the H and with the l_w of the core and the given I, the number of turns is found. The size of wire will be dictated by the current carrying capacity. This figure can be checked by comparing the NA_wF_w of the core checked against the winding volume of the core. From the ratio l_g/l_m and the $_m$ of the core, the gap can be calculated. An example using this approach is described in the Magnetics Catalog and is given in Figure 5.15.

Another example from Snelling (1988)(Figure 5.14) shows a family of curves for different sizes of a particular type of power ferrite core. With the LI^2 known, a horizontal line is drawn until it reaches the first curve and the vertical line from that point gives the NI of the core. The smallest core can be chosen unless other considerations require the use of a larger one. The NI_o is matched against Snelling's table of the NI for the various power cores. Also shown on the Hanna curve is the spacer thickness where the total gap is about $2l_s$. The value of l_g is the gap in the center leg giving the same inductance as $2l_s$. Smith (1983) shows how to use optimization theory to design inductors for minimum resistive loss per unit volume. He has published tables of core geometries which maximize the efficiency of inductors as well as saturable reactors.

Bloom (1989) has described a design method called *Integrated Magnetics* in which the transformer and inductor are combined in a single converter core.

5.7.1-Design of an Inductor for a Switching Regulator

A switching regulator takes an unregulated D.C. output and produces a regulated DC output. In the design of the switching regulator shown in Figure 1.17, an LC filter is used to smooth the ripple in the D.C. output. A typical regulator circuit consists of three parts: transistor switch, diode clamp, and an LC filter. An un-regulated DC voltage is applied to the transistor switch that usually operates at a frequency of 1 to 50 Kilohertz. When the switch is on, the input voltage, E_{in} is applied to the LC filter, thus causing current through the inductor to increase. Excess energy is stored in the inductor and capacitor to maintain output power during the off time of the switch. Regulation is obtained by adjusting the on time, t_{on}, of the transistor switch, using a feedback system from the output. The result is a regulated DC output, expressed as:

$$E_{out} = E_{in}\, t_{on}\, f \qquad [5.57]$$

The off time of the transistor switch is related to the voltages by

$$t_{off} = (1-E_{out}/E_{inmax})/f \qquad [5.58]$$

For E_{inmin}: $$f_{min} = (1-E_{out}/E_{inmin})/t_{off} \qquad [5.59]$$

If we assume the ripple current, i, through the indictor to be equal to $2I_{0min}$, the inductance is;

$$L = E_{out}t_{off}/\, \Delta i \qquad [5.60]$$

The ferrite core to supply this inductance can be obtained by again calculating the LI^2 product and using the charts such as the one shown in Figure 5.16.

From the intersection of the LI^2 with one of the core lines, the appropriate A_L can be read. In this case it is convenient to refer to the standard gapped cores available under each core's description. The number of turns can be calculated from;

$$N = 1000\sqrt{\frac{L}{A_L}} \qquad [5.61.]$$

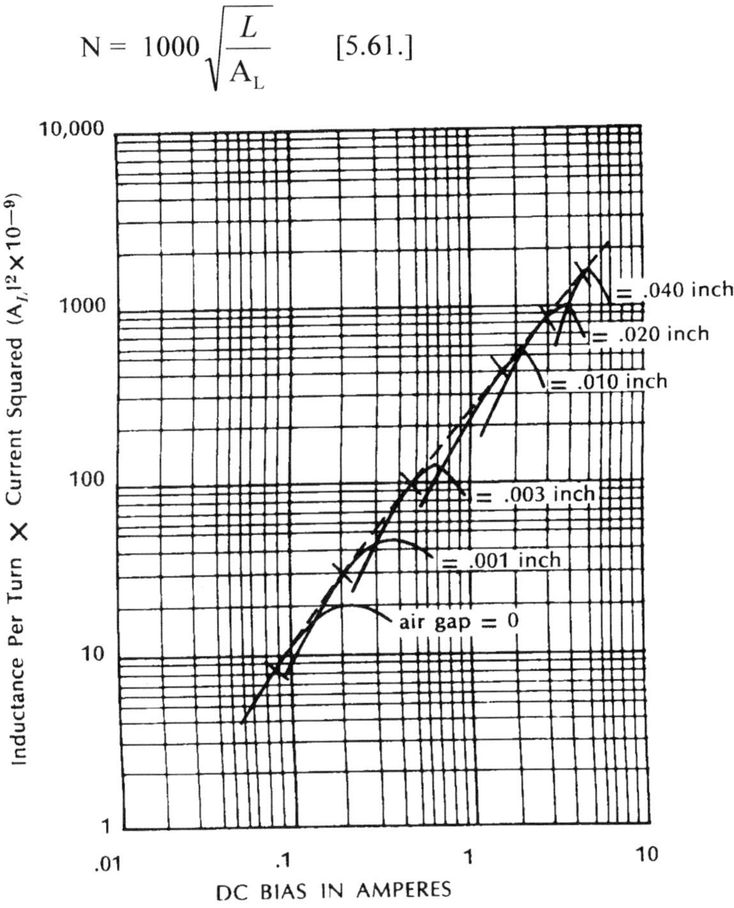

Figure 5.12-A typical Hanna curve . From Fair-Rite Linear Ferrites Catalog(1996)

The wire size is chosen from the wire tables using a current density of 500 circular mils/amp. An example of this method from the Magnetics Catalog is shown in Appendix D.

An approach given by Jongsma (1982a) and contained in the Philips Catalog is shown in Figure 5.16. The LI^2 is plotted against the spacer thickness or center leg gap-width for a series of different core shapes and sizes. When a core supplying the required LI^2 is chosen, reference is made to the data for the individual core chosen.

The specific graph of LI^2 versus spacer thickness for that core is given for various choke designs (depending on the I_{ac}/I_0 ratio). A graph of this type is shown in Figure 5.17. On the same graph, the curve of LI^2 versus A_L for that particular core is given. For the particular LI^2 chosen, the intersection with the line for the converter is found. The working point must be below this line. A vertical can be dropped to the spacer thickness axis and from the tolerance on the spacer thickness, s_{min} and s_{max} can be chosen on the axis. These lines can be extended to the A_L curve for the converter type. The two intersections when read across to A_L (to right hand scale) will give the limits of A_L. To avoid saturation N_{max} is given by;

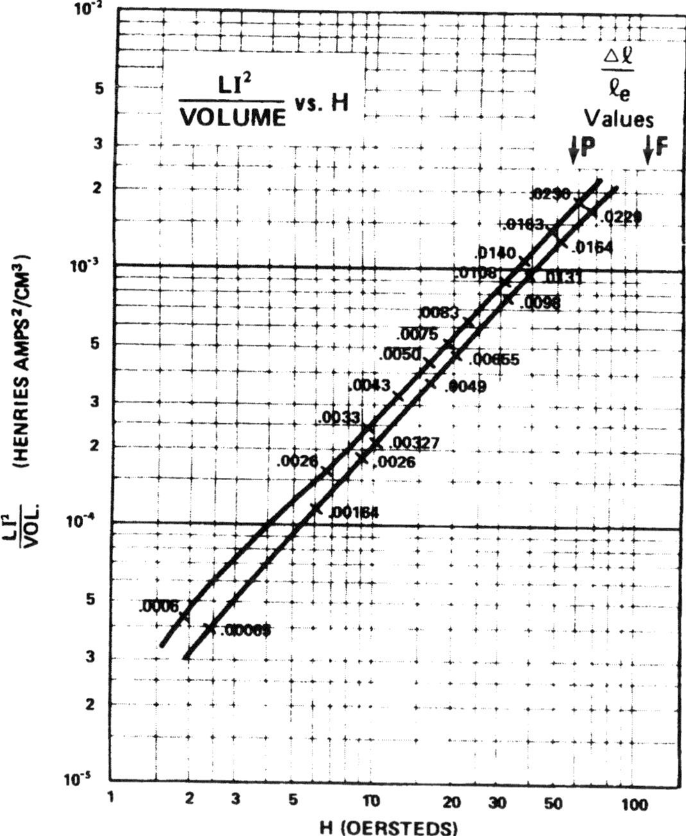

Figure 5.13- A Hanna curve of LI^2/V vs H giving the ratios of gaps to magnetic path lengths to achieve them. From Magnetics Catalog(1989)

$$N_{max} = (I^2L)_{max1}/I_m 2 A_{L1} \qquad [5.62]$$

To achieve L_{min}, the N_{min} is given by;

$$N_{min} = 10^3 \sqrt{\frac{L_{min}}{A_{L2}}} \qquad [5.63]$$

An integral number of turns is chosen. The winding procedure can be completed as outlined under transformers or if special considerations are needed, the design by Jongsma is recommended.

5.7.2-McLyman Treatment of Inductor Design

Following a treatment similar to the one used for transformers, McLyman(1982) employs the K_g constant. The applicable expression is:

$$\alpha = (\text{Energy})^2/K_g K_e \qquad [5.64]$$

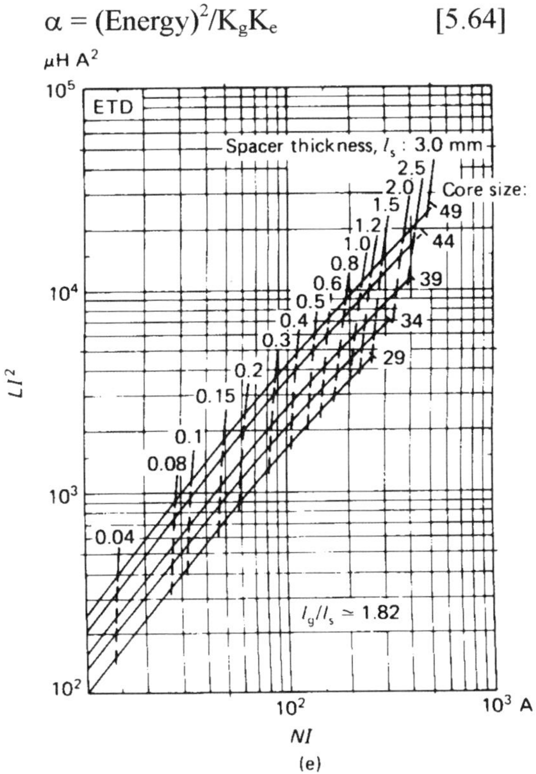

Figure 5.14- A Hanna curve for several ETD core sizes with the spacer thickness required to produce the LI^2 at a particular NI. From Snelling(1988)

where α and K_g have been defined under the transformer calculation. The energy in an inductor is given by

$$\text{Energy} = 1/2\ LI^2 \qquad [5.65]$$

The K_e constant is varied somewhat from the transformer equation. It is represented by:

$$K_e = W_a A_c 2K_u/l_t \qquad [5.66]$$

The area product approach can also be used for inductors:

$$A_p = [2\ (\text{Energy})\ X\ 10^4/B_m K_u K_j]^n \quad cm^4 \qquad [5.67]$$

The fraction, K_u, of the available winding space that will be occupied by the copper is given by

$$K_u = S_1\ x\ S_2\ xS_3\ x\ S_4 \qquad [5.68]$$

where S_1 = conductor area/ wire area
S_2 = wound area/usable window area
S_3 = usable window area/window area
S_4 = usable window area/ usable window area + insulation

The design of an inductor using McLyman's approach is given in Appendix 5.

5.7.3-Flyback Converter Design

Previously, we stated that the design of a flyback converter is similar to that of a power choke or inductor because in both cases, the energy is stored in the inductor during the current rise period and released when the current is turned off. If the converter is a simple non-isolating type (no transformer coupling as shown in Figure 18-6), the design (Jongsma 1982) is treated as a power inductor where;

$$L_{min} = 9\ \delta_{min}\ V_{imax})^2 \text{ and:} \qquad [5.69]$$

$$I_m = I_{dcmax} + 2I_{ac} \qquad [5.70]$$

$$= (\ P_o/\ \delta_{max} V_{imin}) + (\ \delta_{max} V_{imin}/2fL) \qquad [5.71]$$

With the calculated values of I_{max} and L_{min}, the design can then be completed using the power inductor methods. If however, there is transformer coupling as shown in Figure 1.9, the turns ratio must be controlled to avoid damage to the semiconductor switches. Jongsma gives the limiting equations in this case as;

Ferrite DC Bias Core Selector Charts

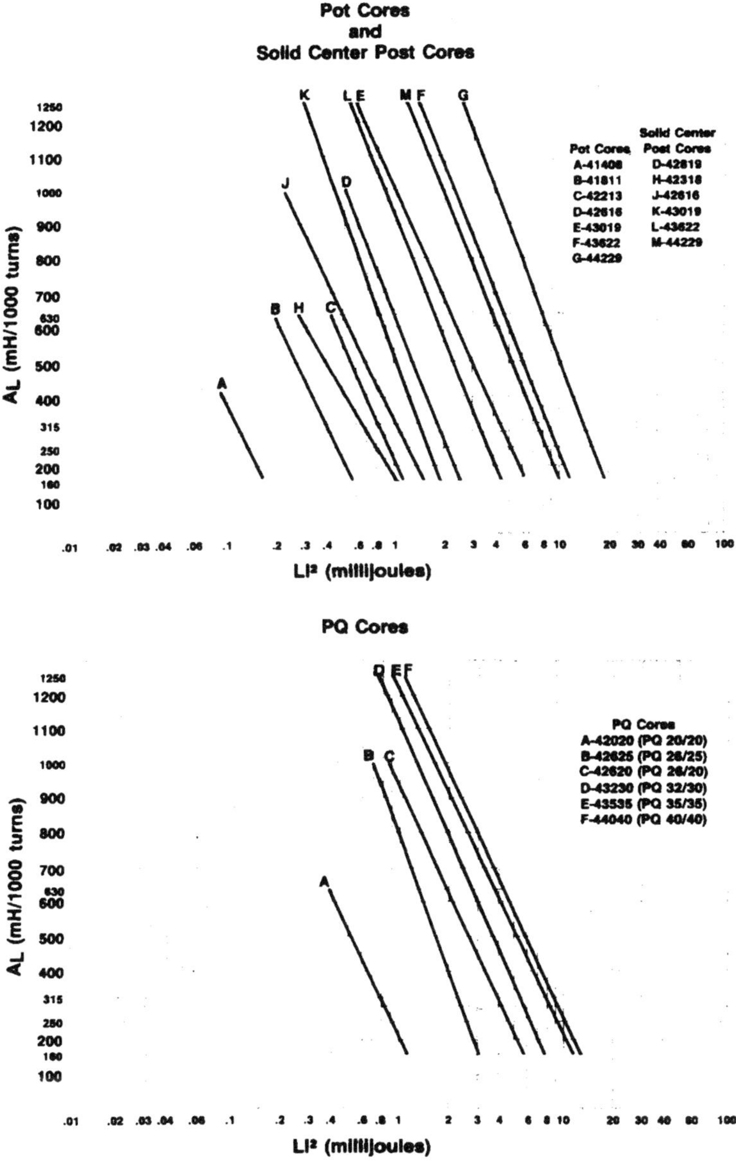

Figure 5.15- A Graph showing the A_L needed in a particular core to furnish a specific LI^2. From Magnetics Catalog (1989)

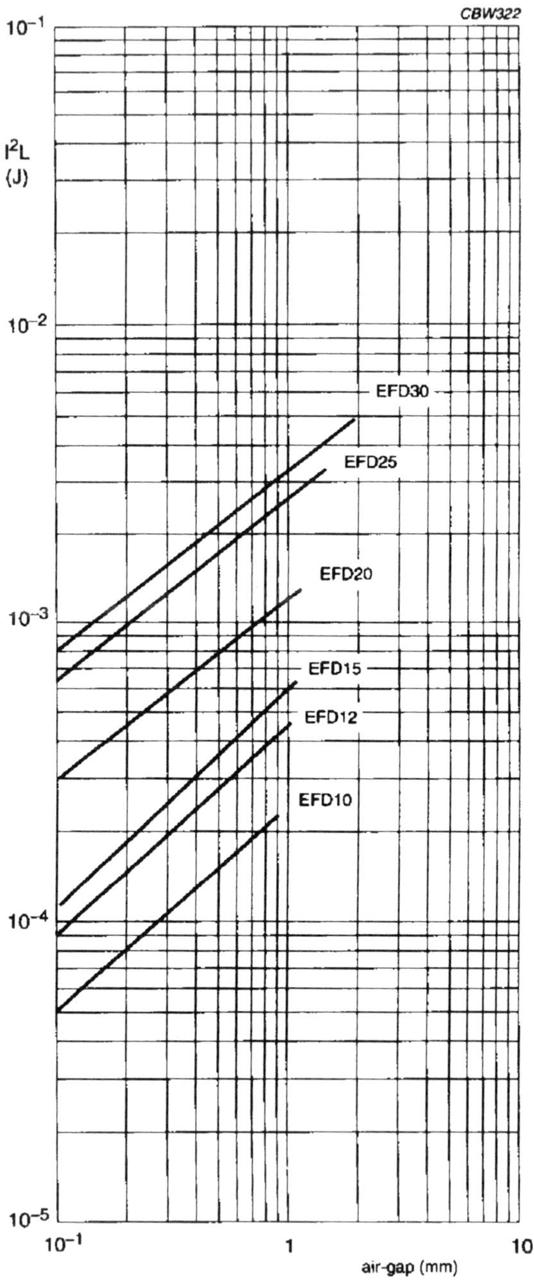

Figure 5.16- The LI^2 for several EFD cores as a function of the center leg gap lengths. From Philip Catalog(1998).

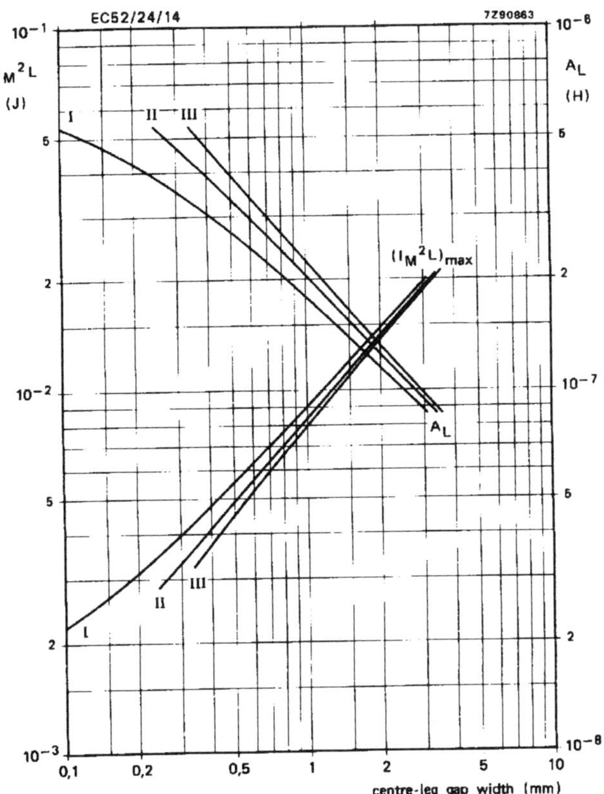

Figure 5.17- For a specific core, the LI^2 and A_L are given for different gap lengths for different converter types. From Philips Catalog(1998)

For $V_{imax}/V_{imin} < 2$

$$r = 3/7\{V_{imax}/V_o + V_F + V_R)\} \qquad [5\text{-}72]$$

Where, V_F = Voltage drop across the output choke

and V_R = Voltage drop across the rectifier

We see then that not only the characteristics of the magnetic devices must be considered but also the voltage drop and current distribution in many of the auxiliary circuit elements. Because of this, when completing the design of these and other magnetic components, the reader is advised to consult the many books, vendors' literature and various periodicals that deal with this subject.

5.8-SWINGING CHOKE

We have spoken of the use of the air gap and the prepolarized cores as techniques in the design of power inductors. Another such design variation that is used to improve the regulation and efficiencies of choke is called the swinging choke or divided gap choke. This is described by Keroes(1969), Martin(1982) and by Snelling(1988). The action is non-linear as shown in Figure 5.18. The use of the stepped gap (Figure 5.19) allows for a wide swing of D.C. currents or magnetic fields. At low D.C. levels, the ripple current is a large part of the total current so a high inductance is needed and is provided by the small gap. However as the D.C. level increases, the ferrite at the small gap will saturate and the large gap will take over, protecting the circuit and main core from saturation and overheating. Thus, a dual action is accomplished.

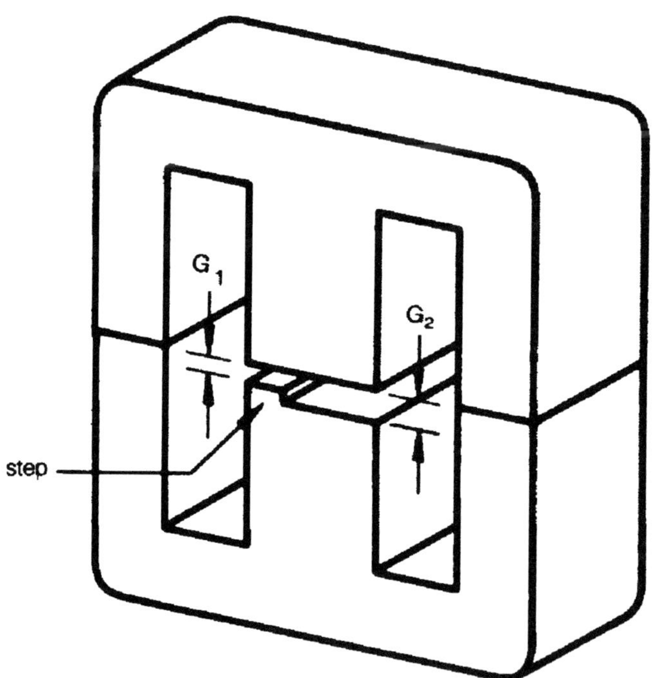

Figure 5.18-The Construction of a swinging-gap choke. From Martin, W.A. Powertechnics Magazine, Feb. (1986) p.19

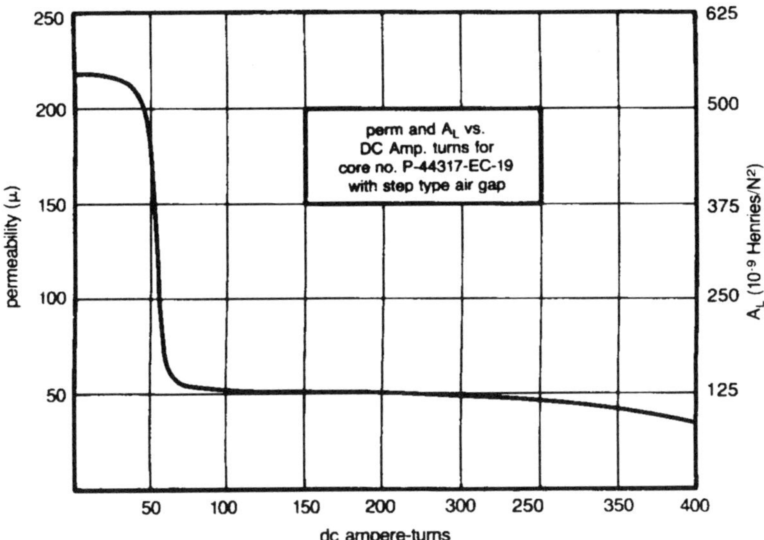

Figure 5.19-The permeability versus D.C. bias characteristics of a swinging-gap choke. From Martin(1986)

5.9-MAGNETIC AMPLIFIER-MULTI-OUTPUT DESIGN

In a multi-output power converter, it is often important to control one or more of the outputs independently. One method of doing this is by using a core having a square-loop material as a magnetic amplifier. As pointed out by Snelling (1976), this delays the leading edge of the secondary circuit 'on' pulse by an amount depending on the re-set condition. The re-set condition determines the amount of volt-seconds needed to drive the core to saturation. When saturation occurs, the inductance falls to a low value and the energy transfer can commence.

Appendix 5.1
Design Example of McLyman K_g Approach

The following section is abstracted from Magnetic Core Selection for Transformers and Inductors by Colonel Wm. T. McLyman, Marcel Dekker, New York,1982

Single-Ended Forward Converter Design

The following parameters are given:

Input voltage (V_{in}) = 140 V min

Output voltage(V_{out})= 10 V

Output current (I_o) = 5.0 A

Frequency(f) = 20 KHz
Switching efficiency= 90%
Regulation = 1.0%
Ferrite toroid matl.= 5000μ (Magnetics)
The design steps used can be summarized as follows

1. Calculate output power which is equal to (V_{out} + $V_{(diode\ drop)}$)x (I_o) = (10 +1)5 = 55 Watts
2. Calculate the apparent power using Equation 5.19
   $P_t = P_0\{\sqrt{2/\eta} + \sqrt{2}\}$
   $= 55\{1.41/.9 + 1.41\} = 164$
   10 % is added to the apparent power for the demagnetizing winding;
   $P_t(1.1) = 180$ Watts
3. Calculate the electrical conditions assuming B_m =.2T and square wave (K_f=4.0) using Equation 5.20
   $K_e = 3712$,
4. Calculate $K_g = P_t/2K_e$, using Equation 5.21
   $K_g = 180/2(3712) = .0242$

   K_g is then recalculated for additional insulation because of the high voltage between primary & secondary windings.
   $K_g = .03025$
5. Select a toroid from McLyman's table (Table 11.5) with the comparable K_g and record the data regarding the toroid.
   Magnetics 52507, $K_g = .0352$
6. Calculate primary turns, N_p, using Equation 2.2 using coefficient 4 for square wave in place of 4.44 for sine and using Teslas for the units for B_m (1T =10^4 Gausses).
   $N_p = V_p \times 10^4 / K_f f B_m A_c$
   $= 140 \times 10^4 / 4 \times .2 \times 2 \times 10^4 \times .393 = 222$
   $N_p = N_m$ (Demagnetizing winding)
7. Calculate primary current, I_p using a duty cycle, D, of .5 and a switching efficiency,η, of .9.

   $I_p = P_0/DV_p\eta = 55/(.5 \times 140 \times .9) = .873$ A.
   $I_m = I_p \times .1 = .0873$ A.

8. Calculate current density, J, from Equation 5.22. Use K_u (Window Utilization Factor) = .4
   $J = 380$
9. Calculate bare wire size A_w(B). For forward converter, I_p and I_m must be multiplied by .707

   $A_{w(B)} = I_p(.707)/J = (.873 \times .707)/380 = .00162$

$A_{w(B)} = I_m(.707)/J = (.0873 \text{ x } .707)/380 = .000162$

10. Select wire size from table
    AWG #25 has bare area of .00162
    $\mu\Omega/\text{cm} = 1062$
11. Calculate primary winding resistance, R_p
    $R_p = (\text{MLT})(\text{N}) \text{ x } \mu\Omega/\text{cm.} = 3.3 \text{ x } 1062 \text{ x}223 \text{ x } 10^{-6}$
    $= .781\ \Omega$

12. Calculate primary copper loss. P_p

    $P_p = (I_p \text{ x } .707\ )^2\ R_p$
    $= (.873 \text{ x } .707)^2 \text{ x } .781 = .297$ Watts
13. Calculate secondary turns, N_s

    $V_s = (\ V_0 + V_d)/D$
    $= (10 + 1)/.9 = 22$
    $N_s = N_p\ V_s/\ V_p$
    $= (223)(22)/140 = 35$ Turns
14. Calculate bare wire size, $A_w(B)$ for secondary

    $A_{w(B)} = I_0(.707)/J$
    $= (5)(.707)/380\ = .00930 \text{ cm}^2$

15. Select wire size from table

    AWG Wire with area of $.00823 \text{ cm}^2$
    $\mu\Omega/\text{cm} = 209.5$

16. Select secondary winding resistance

    $R_s = (\text{MLT})(\text{N})(\mu\Omega/\text{cm})$
    $= (3.3)(35)(209) \text{ x } 10^{-6}\ = .0242\Omega$

17. Calculate secondary copper loss, P_s

    $P_s = (I_0 \text{ x } .707)^2\ R_s$
    $= (5 \text{ x } .707)^2\ .0242 = .302$ Watts
18. Calculate transformer regulation

    $\alpha = P_{cu} \text{ x } 100/(P_0 + P_{cu})$

    where P_{cu} = sum of primary and secondary copper losses
    $= (.297 + .302) = .599$ Watts
    $= (.599 \text{ x } 100)/\ (55 + .599) = 1.08\ \%$
19. Calculate core loss, P_e from core loss curves and core weight.

P_{fe} = (Milliwatts/gm) W_{fe} x 10^{-3}
= (20)(12.1) x 10^{-3} = .242 Watts

20. Calculate Total Losses

$P_\Sigma = P_{cu} + P_{fe}$
= (.599) + (.242) = .841 Watts

21. Calculate efficiency for transformer

$e = (P_o \times 100)/ (P_o + P_\Sigma)$
= [(55) X 100]/(55 + .599) = 98.9%

22. Calculate Watts/Unit Area from surface area of core.

$\Psi = P_\Sigma / A_t$
= (.841)/33.4 = .025 Watts/cm^2

A value of .03 Watts/cm^2 normally gives a 25°C. rise.

APPENDIX 5.2

The following section is abstracted from the Magnetics Catalog FC405, published by Magnetic, Division of Spang and Co., Butler PA 16001, 1987

Magnetics Inductor Design Method using Hanna Curves
Example - The following example illustrates the use of a Hanna curve to find the core for a particular power inductor.

Let L = .1 mH and I_{DC} = 10 amperes. Find the core, the air gap and number of turns required.

1. Calculate LI^2
$LI^2 = (.1 \times 10^{-3}) \times (10)^2 = 10 \times 10^{-3}$

2. Refer to Hanna Curve in Figure 5.16. Assume $(LI)^2/V = 5 \times 10^{-4}$ (from center of vertical scale).

3. Core Selection- Choose a core geometry, for example an E core, and select a size with the volume nearest to 20 cm^3. Use P45021-EC.

Volume = 21.6 cm^3,l_e=9.58cm,W_a=.351 x 10^6 circ.mils

4. Recalculate

$LI^2/V = 10 \times 10^{-3}/21.4 = 4.6 \times 10^{-4}$

5. Determine H and l_g/l_e from the Hanna curve (P material),using recalculated value of LI^2/V.

H=18 and $l_g/l_e = .006$

6. Calculate N

H = .4 NI/l, N = Hl/.4 I = 13.7Turns-Use N=14

7. Calculate W_a needed. For I_{DC} = 10 amperes, use AWG #11 wire.

W_a = 9 x 10^3 cir. mils per turn

W_a needed = A_w x N/K

where W_a = core or bobbin window area

Aw = cross sectional area of the wire

N = number of turns

K=winding (or space utilization) factor (K varies with the designer and operating conditions of the inductor. Typically, this factor is 0.4).

Wa needed = (9 x 10^3) x (14/0.4) = 315 x 10^3 circ. mils

8. Compare Wa values

W_a needed = 315 x 10^3 circ. mils

W_a available in 45021-EC = 351 x 10^3 circ. mils

At this point, the designer can use the core selected or repeat this process to select a smaller (or larger) core.

9. Gap calculation. If the P-45021-EC core is chosen, the air gap is calculated as follows.

$l_g/l_e = .006$, $l_e = 9.58$ cm.

$l_g = .006 \times 9.58 = .057$ cm.(.023 in.)

APPENDIX 5.3

Magnetics Inductor Design for Switching Regulators

The following is abstracted from Magnetics Catalog FC405, published by Magnetics, Division of Spang & Co.,Butler, PA 16001, 1987

Only two parameters of the design application must be known:

(a) Inductance required with DC bias

(b) DC current

1. Compute the product of LI^2 where:
   L = inductance required with DC bias (millihenries)
   I = maximum DC output current = I_{omax} + i

2. Locate the LI^2 value on the Ferrite Core Selector charts such as the one shown in Figures 5.16. Follow the LI^2 coordinate to the intersection with the first core size curve. Read the maximum nominal inductance, A_L, on the Y axis. This represents the smallest core size and maximum A_L at which saturation will be avoided.

3. Any core size line that intersects the LI^2 coordinate represents a workable core for the inductor if the core's A_L value is less than the maximum value obtained on the chart. If possible, it is advisable to use the standard gapped cores because of their availability. These are indicated by dotted lines on the charts and can be found in the catalog.

4. Required inductance L, core size, and core nominal inductance (A_L) are known. Calculate the number of turns using

$$N = 10^3 \sqrt{\frac{L}{A_L}}$$

where L is in millihenries.

5. Choose the wire size from the wire tables using 500 circular mils per amp.

Example - Choose a core for a switching regulator with the following requirements:

E_0 = 5 Volts
e_o = .5 Volts
I_{omax} = 6 amp
I_{omin} = 1 amp
E_{inmi}n= 25 Volts
E_{inmax}= 35 Volts
f = 20 KHz.

1. Calculate the off-time and minimum switching, f_{min}, of the transistor switch using equations 5.58 and 5.59.

$$t_{off} = (1-E_{out}/E_{inmax})/f$$
$$t_{off} = (1-5/35)/(20{,}000) = 4.3 \times 10^{-5} \text{ sec.}$$
$$f_{min} = (1-E_{out}/E_{inmin})t_{off}$$

$f_{min} = (1\text{-}5/25)/(4.3 \times 10^{-5}) = 18{,}700$ Hz.

2. Let the maximum ripple current, i, through the inductor be

$\Delta i = 2I_{omin}$
$\Delta i = 2(1) = 2$ Amps

3. Calculate L using Equation 5.60.

$L = (E_{out} \times t_{off})/\Delta i$
$L = 5(4.3 \times 10^{-5})/2 = .107$ millihenries

4. Calculate the value of the capacitance, C and maximum equivalent series resistance, ESR max

$C = i/8f_{min}\,\Delta e_o$
$C = 2/8(18700)(.5) = 26.7\ \mu$ farads

$ESR_{max} = \Delta e_o / \Delta i$
$ESR_{max} = .5/2 = .25$ ohms

5. The product of $LI^2 = (.107)\,(8)^2 = 6.9$ millijoules.

6. Due to the many shapes available in ferrites, there can be several choices for the selection. Any core size that the LI^2 coordinate intersects can be used if the maximum A_L is not exceeded. Following the LI^2 coordinate, the choices are:

(a) 45224 EC 52 core, A_L315
(b) 45015 E core, A_L250
(c) 44229 solid center post core, A_L315
(d) 43622 pot core, A_L400
(e) 43230 PQ core, A_L250

7. Given the A_L, the number of turns the required inductance can be found for each core using Equation 5.61.

A_L	Turns
250	21
315	19
400	17

8. Use #14 wire.

APPENDIX 5.4

MCLYMAN DESIGN -SWITCHING INDUCTOR - K_G APPROACH

This section is abstracted from Magnetic Core Selection for Transformers and Inductors by Colonel Wm. T. McLyman, Marcel Dekker, New York, 1982

Design of a Buck Switching Inductor

Given;

Input Voltage, V_i = 28 +/- 6 V.

Output Voltage, V_o = 20 V.

Output Current Range, I_0 = 5 - 0.5 A

Frequency, f, = 20 KHz.

Switching Efficiency, = 98%

Regulation = 1.0%

Ferrite Pot Core

Step 1. Calculate time period, t, of operation

$t = 1/f = 1/20\text{x } 10^3 = 50 \text{ x } 10^{-6}$ s.

Step 2. Calculate Minimum duty cycle

$D_{min} = V_0/V_{in(max)} = 20/34(0.98) = .60$

Step 3 Calculate maximum duty cycle

$D_{max} = V_0/V_{in(min)} = 20/22(0.98) = .927$

Step 4. Calculate Load Resistance at Minimum Load Current

$R_0 = V_0/I_{0(min)} = 20/0.5 = 40$

Step 5. Calculate Minimum Required Inductance

For a Buck Converter

$L_{min} = R_{0(min)}t(1\text{-}D)_{min})/2 = (40)(50\text{x}10^{-6})(1\text{-}0.6)/2$

$= 400 \text{ x } 10^{-6}$ H.

Step 6. Calculate ΔI in the Inductor

$\Delta I = tV_{in(max)}D_{(min}(1\text{-}D_{(min)})/L$

$= (50 \text{ x}10^{-6})(34)(0.6)(1\text{-}0.6)/400 \text{ x } 10^{-6}$

$= 1.0 = 2I_{0(min)}$ A.

Step 7. Calculate $LI^2/2$

$I = I_{0(max)} + \Delta I/2 = (5.0) + 1.0/2 = 5.5$ A.

$LI^2/2 = (400 \text{ x } 10^{-6})(5.5)^2/2 = .00605$ W-s.

Step 8. Calculate K_e using Equation 5.66.

$P_0 = V_0I_0 = (20)(5) = 100$ W.

Assume B_m = .35 T (3500 Gausses)

$K_e = 0.145\ P_0B^2\ x10^{-4} = 0.000178$

Step 9. Calculate K_G

K_G =(Energy)^2/K_e using Equation 5.64

= $(0.00605)^2/(0.0001781) = 0.213$

Step 10. Select a comparable core geometry K_g from listing of pot cores (McLyman, 1982). Record all pertinent dimensional data.

Pot core = B65611, 36x22(Siemens)

For this pot core, K_g = 0.221

G(window height) =1.46cm.

W_{tfe}(weight ferrite) = 26 gm.

MLT = 7.3 cm.,A_c = 2.01 cm^2, W_a = 1.00 cm^2

A_t(Surface area) = 45.24 cm^2

Step 11. Calculate current density (Correlation with K_g is derived in book (McLyman 1982)

$J = 2(\text{Energy})\ x\ 10^4/B_mK_uA_p$

Use K = 0.4 (window utilization factor)

J= 2(0.00605) x 10R4F/(0.35)(0.4)(2.01) = 430A/cm^2

Step 12. Calculate bare wire size $A_{w(B)}$

$A_{w(B)} = I_0/J = 5/430 = .0116$

Step 13. Select a wire size from the wire table. If area is not within 10%,take smaller size. Record wire data.

AWG #17 with 0.01039 cm^2

μΩ/cm = 166

A_w = 0.0117 cm^2 (with insulation)

Step 14. Calculate the effective window area, $W_{a(eff)}$

$W_{a(eff)} = W_aS_3$

For single section bobbin on pot core, S_3 = 0.75

$W_{a(eff)}$ = (1.00) (0.75) = 0.75

Step 15. Calculate number of turns

$N = W_{a(eff)}\ S_2/A_w$

Typical value of S_2 = 0.6

N = (0.75)(0.6)/(0.0117) = 38 turns

Step 16. Calculate gap required for inductance

$l_g = 0.4\ N^2A_c\ x10^{-8}/\ L$

$= (1.26)(38)^2(2.01)\ x\ 10^{-8}/412\ x10^{-6} = 0.089$ cm.

For fishpaper spacer, thickness is given in mils.

$l_g = 0.089$ cm x 393.7 mils/cm = 35 mils

Paper comes in 10 and 7 mils .One of each across entire pot core mating surface doubles gap.(a gap each for skirt and centerpost) Therefore the total gap is 17 mils or 0.034x2.34 = 0.0864cm.

Step 17-18. Recalculate new turns correcting for fringing flux (not shown here)

N =34

Step 19. Calculate Winding Resistance

$R = (MLT)(N)\ \Omega\ /cm \times 10^{-6}$
$= (7.3)(34)(166) \times 10^{-6}$
$= 0.041\Omega$

Step 20. Calculate Copper Loss, P_{cu}

$P_{cu} = I^2R = (5.5)^2(0.041) = 1.24$ W.

Step 21. Calculate Regulation, .

$\alpha = P_{cu} \times 100/(P_0 + P_{cu}) = (1.24)(100)/(100+1.24)$
$= 1.22\%$

Step 22. Calculate total a.c. + d.c. flux density

$B_m = 0.4\pi N(I_{dc} + \Delta I/2) \times 10^{-4}/l_g$
$= (1.26)(34)(5.5) \times 10^{-4}/0.0864 = 0.273T(2730G.)$

Step 23. Calculate a.c. flux density

$B_{mac} = 0.4\pi N(\Delta I/2) \times 10^{-4}/l_g$
$= (1.26)(34)(0.5) \times 10^{-4}/\ 0.0864 = 0.0248$ T

Step 24. Calculate Core Loss P_{fe}. Use core loss curves for 0.0248 T. or 248 Gausses. Use the ferrite weight given before.

$P_{fe} = (mW/gm)(W_{tfe}) \times 10^{-3}$
$= (0.06)(57) \times 10^{-3} = 0.0034$ W.

Step 25. Calculate total loss

$P_t = P_{cu} + P_{fe} = (1.24) + (0.0034) = 1.2434$ W.

Step 26. Calculate the efficiency

$e = (P_0)(100)/(P_0 + P_t) = (100)(100)/(100+1.2434)$
$= 98.8\%$

Step 27 Calculate the Watts/unit area

$\psi\ = P\ \ /A_t = (1.2434)/45.24 = 0.0275\ W/cm^2$

The value of .03 W/cm^2 corresponds to a temperature rise of 25°C.

APPENDIX 5.5

Design of Output Inductor using Metglas Amorphous Choke Cores

The following is the design procedure suggested by Honeywell (previously Allied-Signal) for the design of a high frequency output inductor using Metglas amorphous choke cores.

1. Determine the Choke Ripple Current, ΔI- For continuous operation, the minimum DC current I_{omin}, must be equal to or greater than 1/2 the choke ripple current, ΔI

$$I_{omin} \;=/> \Delta I/2 \qquad \Delta I =/< 2\, I_{omin}$$

When the minimum DC current is not given, assume that ΔI is 10-20% of I_{omax}.

$$\Delta I = 0.2\, I_{omax}$$

2. Determine the critical inductance, L_{min}

$$L_{min} = E_o(1 - D_{min})/(2 \times I_{omin} \times f)$$

Where $D_{min} = E_o/(E_o - E_{pk})$
and $t_{offmax} = 1 - D_{min}$

$$T = t_{on} + t_{off} = 1/f$$

$$t_{on} = DT \text{ and } t_{off} = T - DT$$

For average volt-seconds across inductor to equal zero,

$$DT\, E_{pk} = E_o\ (T - DT)$$

Solving for D;

$$D = E_o/\ (E_o + E_{pk})$$

3. Determine the required energy;

$$I_{pk} = I_{omax} + (\Delta I/2)$$

$$W = L_{min}\, I_{pk}^2$$

4. Determine the area product W_aA_c;

$W_aA_c = \{2W /(B_{max} KJ)\} \times 10^4 \ (cm^4)$

Choose a core from the the Metglas Core Table

5. Calculate the number of turns and wire size

$N = \{L_{min} (nH)/A_L\}^{0.5}$

$A_W = KW_A/ N$

6. Calculate the ripple current, $B_{\Delta I}$

$B_{\Delta I} = E_{pk} t_{on}/A_cN$

7. Calculate the DC flux density, B_{DC}

$B_{DC} = 0.4 \pi NI_{DC} \times 10^4/l_m$

Check for the maximum flux density, B_{max}

$B_{max} = B_{DC} + B_{\Delta I}/2$ Teslas

Check for percent permeability vs DC Bias(From μ vs DC Graph)

$H = 0.4 \pi NI_{DC}/l_m$ Oersteds

8 . Losses and Temperature Rise

$B_{ac} = B_{\Delta I} /2 = E_{pk} t_{on}/2A_cN$

$P_c(W/Kg) = (87.95) (B_{ac})^{2.01}(f)^{1.29}$

$P_c(W) = P_c(W/Kg) \times wt.(Kg)$

Calculate copper loss

$R = (MLT) \times (R_w) \times (N)$

$P_w = (I_o)^2 R$

Calculate Total Loss

$$P = P_c + P_w$$

Estimate Temperature Rise

$$SA = \{\pi(OD)_{wound}{}^2/2\} + \{\pi(OD)_{wound} \times (ht_{core} + OD_{wound} + OD_{core})$$

Appendix 5.6
Push-Pull Output Inductor Design Using a LPT E2000Q Core
Nanocrystalline Core by Coremaster International

Article AN114 by Colonel Wm T. McLyman

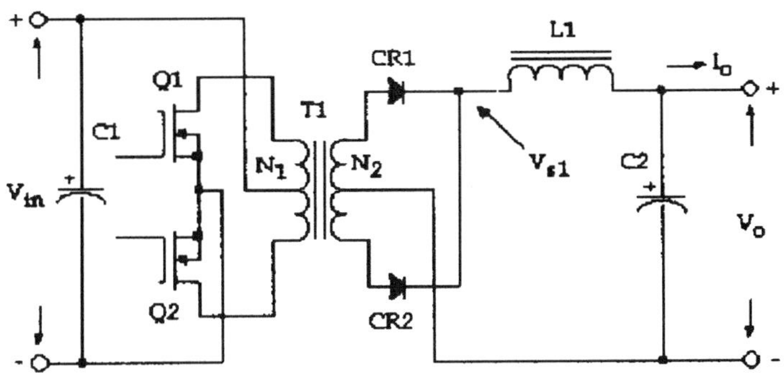

Figure 5.20-Push-Pull Converter with Single Output

1. Frequency	$f = 100$ KHz.
2. Output Voltage	$V_o = 5$ V.
3. Output current, max	$I_{0max} = 10$ A
4. Output current, min	$I_{0min} = 2$ A
5. Delta current	$\Delta I = 4$A
6. Input Voltage, maximum	$V_{simax} = 9$V
7. Input Voltage, minimum	$V_{simin} = 6$V
8. Regulation	$\alpha = 1.0\%$
9. Output Power	$P_0 = 50$ W.
10.Operating Flux Density	$B_M = 0.8$ T
11.Window Utilization	$K_U = 0.4$
12.Diode Voltage Drop	$V_d = 1$ V

Step No. 1. Calculate the period, T

$$T = 1/f$$

$$T = 1/100{,}000 = 10.\ 10^{-6}$$

Step No. 2.Calculate the minimum duty ratio, D_{min}

$$D_{min} = V_0/V_{max}$$

$$D_{min} = 5/9 = 0.555$$

Step No. 3. Calculate the required inductance, L

$$L = T.(V_0 - V_d).(1 - D_{min})$$
$$L = 10.\ 10^{-6}.\ 6.\ (1\text{-}0.555) = 6{,}675 \text{ use } 7\ [\mu H]$$

Step No.4. Calculate the peak current, I_{pk}

$$I_{pk} = I_{0max} + \Delta I/2$$
$$I_{pk} = 10 + 4/2 = 12 \text{ A}$$

Step No.5. Calculate the energy-handling capability in watt-seconds, [w.s]

$$\textit{Energy} = L.I_{pk}^{\ 2}/2$$
$$\textit{Energy} = 7.\ 10^{-6}.12^2 = 0.000504 \text{ [w.s]}$$

Step No.6. Calculate the electrical condition, K_e

$$K_e = 0.145.\ P_0.\ B_M^{\ 2}.\ 10^{-4}$$
$$K_e = 0.145.\ 50.\ 0.8^2.\ 10^{-4} = 0.000464$$

Step No.7. Calculate the core geometry, K_g

$$K_g = \textit{Energy}^2/\ K_e.\ \alpha$$
$$K_g = 0.000504^2/0.000464.1.o = 0.000547 \text{ [cm}^2$$

Step No.8. Select from the LPT data sheet a E2000Q core comparable in core geometry, K_g

Core number	GC70111
Manufacturer	CMI
Magnetic path length, MPL	4.1 cm
Core weight, W_{tfe}	4.3 g.
Copper weight, W_{cu}	5.6 g.
Mean length per turn (MLT)	2.7 cm.
Iron area, A_c	0.14 cm^2
Window Area, W_a	0.581
Area Product, A_p	0.08132 cm^4

Core geometry, K_g	0.00168 cm^5
Surface Area, A_t	16.3 cm^2
Permeability	$\mu = 300$
MilliHenries per 1000 turns	mH= 129

Step No. 9.Calculate the rms current

$$I_{rms} = \sqrt{I_{max}^2 + (\Delta I/2)^2}$$

$$I_{rms} = \sqrt{10^2 + 2^2} = 10.2\ [A]$$

Step No.10.Calculate the current density, J,using a window utilization,K_u= 0.4

$$J = (2.\ \mathit{Energy}.10^4)/A_p.\ B_m.K_u$$
$$J = (2 \text{ x } 0.000504 \text{ x } 10^4)/(0.08132 \text{ x } 0.8 \text{ x}0.4) = 387\ [A/cm^2)$$

Step No.11 Calculate the required permeability, $\Delta\mu$

$$\Delta\mu = (B_m.MPL.\ 10^4)/(0.4\pi.W_a.J.\ K_u)$$

$$\Delta\mu = (0.8 \text{ x } 3.14 \text{ x } 0.581 \text{ x}.387 \text{ x } 0.4) = 290 \text{ use } 300$$

Step No.12. Calculate the number of turns

$$N = 1000\sqrt{L_{(new)} / L_{1000}}$$

$$N = 1000\sqrt{0.007/129} = 7.37 \text{ use 7 turns}$$

Step No. 13 Calculate the peak flux density, B_m

$$B_m = (0.4\pi N I_{pk}\ \mu_\Sigma.10^4)/MPL$$
$$B_m = (0.4 \text{ x}3.14\text{x } 7 \text{ x } 12\text{x } 300.\ 10^4)/4.06 = 0.779\ [T]$$

Step No.14. Calculate the required bare wire area, $A_{w(B)}$.

$$A_{w(B)} = I_{rms}/J$$
$$A_{w(B)} = 10.2/387 = 0.0264\ [cm^2]$$

Step No.15. Select the wire size with the required area from the Wire Table. If the area is not within 10% of the required area, then go to the next smaller size.

$$AWG = 13$$
$$A_{w(B)} = 0.0263[cm^2]$$
$$\mu\Omega/cm = 65.5$$

Step No. 16. Check the ΔI current density using the skin effect ε.

$$s = 6.62/\sqrt{f}\ \sqrt{100{,}000}$$
$$s = 6.62/\sqrt{100{,}000} = 0.0209\ [cm]$$

Calculate the diameter of a #13 AWG

$$D = \sqrt{4A_{w(B)}/\pi}$$
$$D = \sqrt{4x0.0263/3.14} = 0.183\ [cm]$$

Subtract 2 times the skin depth from the diameter and calculate the new area.

$$D_n = D\text{-}2\varepsilon$$
$$D_n = 0.183 - 2x\ 0.0209 = 0.141\ \ [cm]$$
$$A_n = \pi D_n^2/4$$
$$A_n = 3.14\ x0.141^2/4 = 0.0156\ [cm^2]$$

Take the difference between $A_{w(B)}$ and A_n. This will be the area for the ΔI current.

$$A_{\Delta I} = A_{w(B)} - A_n$$
$$A_{\Delta I} = 0.0263 - 0.0156 = 0.0107\ [cm^2]$$

Check the current density to see if it is close to the designcurrent density, J.

$$J = \Delta I/\ A_{\Delta I}$$
$$J = 4/0.0107 = 374\ \ [A/cm^2]$$
$$\Delta I \text{ current density} = 374$$
$$\text{DC current density} = 387$$

Step No.17.Calcuate the winding resistance, R

$$R = MLT\ xN\ x\ (\mu\Omega/cm)\ x10^{-6}$$
$$R_p = 2.7\ x\ 7\ x\ 65.6\ x10^{-6} = 0.00124\ [\Omega]$$

Step No. 18. Calculate the copper loss, P_{cu}

$$P_{cu} = I_{rms}^{\ 2} R$$
$$P_{cu} = 10.2^2 \text{ x } 0.00124 = 0.124 = 0.129 \text{ [W]}$$

Step No. 19. Calculate the magnetizing force in Oersteds, H

$$H = 0.4\pi \ NI_{pk}/MPL$$
$$H = 0.4 \text{ x } 3.14 \text{ x } 7\text{x } 12/4.06 = 25.98 \text{ [Oe]}$$

Step No. 20. Calculate the ac flux density in T, B_{ac}

$$B_{ac} = (0.4\pi N \text{ x } \Delta I/2)\text{x } \mu_r \text{ x}10^4/MPL$$
$$B_{ac} = 0.4\pi \text{ x } 7\text{x } 2 \text{ x } 300 \text{ x } \ 10^4/4.06 = 0.13 \text{ [T]}$$

Step No. 21 Calculate the regulation, α, for this design

$$\text{New regulation} = K_{g(required)}/K_{g(used)} = 0.00547/0.00168 = 0.326$$

Step No. 22. Calculate the Watts/Kilogram, W/K

$$W/K = 8.64 \text{ x } 10^{-7} \ f^{\,1.834} \text{ x } B_{ac}^{\ 2.112}$$
$$W/K = 8.64 \text{ x } 10^{-7\,x}\ 100{,}000^{1.834} \text{ x } 0.13^{2.112} = 17.0 \text{ [W]}$$

Step No. 23. Calculate the core loss, P_{fe}

$$P_{fe} = (mW/g) \ W_{fe} \text{ x}10^{-3}$$
$$P_{fe} = 17 \text{ x } 4.3 \text{ x}10^{-3} = 0.0738 \text{ [W]}$$

Step No. 24. Calculate the total loss, P_Σ

$$P_\Sigma = P_{cu} + P_{fe}$$
$$P_\Sigma = 0.129 \ 0.0731 \ 0.202 \text{ [W]}$$

Step No. 25. Calculate the watt density, Ψ

$$\Psi = P_\Sigma / A_t$$
$$\Psi = 0.202/16.3 = 0.0124 \text{ [W/cm}^2\text{]}$$

Step No. 26. Calculate the temperature rise, T_r

$$T_r = 450 \times \Psi^{0.826}$$
$$T_r = 450 \times 0.0124^{0.826} = 1.98\ [^{o}C.]$$

Step No. 27. Calculate the window utilization factor , K_U

$$K_U = N\ S_N\ A_{w(B)}$$
$$K_{U} = 7 \times 1 \times 0.0263/\ 0.581 = 0.317$$

References

Bracke, L.P.M.,(1983) Electronic Components and Applications, Vol.5, #3 June 1983,p171

Bracke L.P.M.(1982) and Geerlings, F.C., High Frequency Power Transformer and Choke Design, Part 1, NV Philips Gloeilampenfabrieken, Eindhoven, Netherlands

Buthker,C.(1986) and Harper, D.J., Transactions HFPC, 1986,186

Carsten, B.(1986), PCIM,Nov.1986,34

De Maw,M.F.(1981),Ferromagnetic core Design and Applications Handbook, Prentice Hall, Englewood Cliffs, NJ, 1981

Grossner, N.R.(1983), Transformers for Electronic Circuits, McGraw-Hill Book Co., New York

Hanna,C.R.,J.Am.(1927) I.E.E., 46,128,

Hess, J.(1985) and Zenger, M., Advances in Ceramics, Vol.16 501

Hiramatsu, R.(1983) and Mullett, C.E.,Proc. Powercon 10,F2, 1

Hnatek,E.R.(1981), Design of Solid State Power Supplies, Van Nostrand Reinhold,New York

IEC (19) Document 435,International Electrotechnical Commission

Jongsma, J.(1982),High Frequency Ferrite Power Transformer and Choke Design, Part 3, Pilips Gloeilampenfabrieken, Eindhoven Netherlands

Jongsma, J.(1982a) and Bracke, L.P.M. ibid Part 4

Magnetics (2000) Ferrite Core Catalog, Magnetic Div.,Spang and Co, Butler, PA 16001

Magnetics (1984) Bulletin on Materials for SMPS

Martin,H.,(1984), Proc. Powercon 11, B1, 1

Martin, W.A.(1978), Electronic Design, April 12,1978, 94

Martin, W.A.(1986), Powertechnics Magazine,Feb.1986,p.19

Martin, W.A.(1982), Proc. Powercon 9

Martin, W.A.(1987), Proceedings,Power Electronics Conference (1987)

McLyman, Col. W.T.(1969) JPL, Cal Inst. Tech. Report 2688-2

McLyman, Col.W.T.,(1982), Transformer and Inductor Design Handbook, Marcel Dekker, New York

McLyman, Col.W.T.(1982), Magnetic Core Selection for Transformers and Inductors, Marcel Dekker, New York
McLyman, Col. W.T. (1990) KG Magnetics Magnetic Component Design Software Program
Pressman, A.(1977) Switching and Linear Power Supply Converter Design, Hayden Book Co., Rochelle Park, N.J.
Philip Catalog,(1986) Book C5, Philips Components and Materials Div., 5600Md, Eindhoven, Netherlands
Roddam, T.(1963), Transistor Inverters and Converters, Iliffe, London and Van Nostrand Reinhold, New York
Siemens (1986-7) Ferrites Data Book, Siemens AG, Bereich Bauelemente, Balanstrasse 73, 8000 Munich 80 Germany
Smith, S.(1983), Magnetic Components, Van Nostrand Reinhold, New York
Smith, S. (1983a) Power Conversion International, May 1983, 22
Snelling E.(1988) Soft Ferrites, Properties and Applications Butterworths, London
Snelling, E(1989) presented at ICF5
Stijntjes,T.G.W.(1985), and Roelofsma, J.J., Advances in Ceramics, Vol 16, 493
Stijntjes, T.G.W.(1989) Presented at ICF5, Paper C1-01
TDK (1988) Catalog BLE-001F, June 1988, TDK, 13-1 Nihonbashi, Chuo-ku, Tokyo, 103, Japan
Thomson (1988) Soft Ferrites Catalog, Thomson LCC, Courbeville, Cedex, France
VDE ( ) Document 0806
Watson, J.K.(1980) Applications of Magnetism, John Wiley and Sons, New York
Watson, J.K.(1986) IEEE Trans Magnetics
Wood, P.(1981) Switching Power Converters, Van Nostrand Reinhold, New York
Zenger,M. (1984), Proceedings, Powercon 11(1984)

APPENDIX 5.6
RECENT ARTICLES ON DESIGN OF FERRITES FOR POWER APPLICATIONS

Baasch, T.L., Electronic Products, Oct. 1971, 25
Bledsoe, C., Electronic Business, June 1,1984, 128
Bloom,E., IEEE Transactions on Magnetics,(1986), 141
Bosley, L.M. (1994) Magnetics Brochure.
Brown, B.(1992) PCIM, July 1992) 46
Brown, J.F., Powetechnics. Dec.1986, 17
Carlisle, B.H.(1985) Machine Design,Sept. 12, 53

Cattermole,P.(1988) and Cohn,Z..Proc. HFPC,1986, 111
Chen, D.Y., Solid State Power Conversion, Nov/Dec.,1978,50
Ciarcia, S.A.,Byte,Nov.1981, 36
Cuk,S., Power Conversion International, 1981, 22
Dull,W., Kusko, A.& Knutrud,T., EDN, Mar.5,1975, 47
Engelman, R.(1989)PCIM, 15, #7, 14
Estrov, A.(1989) PCIM, May, 1989 16
Estrov, A.(1986) PCIM, August 1986, 14
Finger, C.W. (1986) Power Conversion International,1986
Fluke,J.C.,Proc. Power Electronics Show, 1986, 128
Gatres B.(1992) PCIM, 18, #7 July 1992, 28
Harada, H. and Sakamoto, K., IEEE Translation Journal of Magnetics in Japan, #7, Oct.1985
Hew, E., Power Conversion International, Jul./Aug. 1982, 14
Hill, P.C., Proc. Powercon 2, 1975, 243
Hiramatsu, R. (1983),and Mullett, C.E.,Proc. Powercon,10, F2,1
Kamada, . (1985), and Suzuki, K., Advances in Ceramics,Vol. 16,507
Kepco (1986) Kepco Currents,Vol 1.#2
Kitagawa, T. and Mitsui, T.,IEEE Translation Journal of Magnetics in Japan, Sept, 1985
Konopinski, T. and Szuba, S. Electronic Design, 12,June 7, 1979, 86
Margolin, B., Electronic Products, Mar.28,1983, 53
Martin, H. (1984) Proc. Powercon,11, B1,1
Martin, W.A. Proc. Powercon 9, 1982,
Middlebrook, R.D., Power Conversion International, Sept. 1983,20
Mochizuki, T. (1985) Sasaki, I. And Torii, M.,Advances in Ceramics, Vol. 16,487
Mohandes B.E.(1994) PCIM, July 1994, 8
Mullett, C.E.,Proc. Power Electronics Show,1986,36
Sano, T. (1988), Morita, A. and Matsukawa, A. Proc. PCIM,July 1988, 19
Schlotterbeck, M.(1981),and Zenger,M., Proc. PCIM,1981,37
Shiraki, S.F.(1980) Proc. Powercon,7,J4,1
Smith, S., Power Conversion International, May 1983, 22
Stratford, J.M., EDN, Oct. 13,1983, 140
Sum, K.K.,Power Electronics,1986, 153
Triner, J.E.,Power Conversion International, Jan. 1981, 69
Turnbull, J., Electronic Products, May,15,1972, 53
Ying X. and Zhi,Z.,IEEE Transactions on Magnetics,Feb.1985,148
Zenger,M.(1984) Proc. Powercon,11(1984

Chapter 6
COMMERCIALLY-AVAILABLE COMPONENTS FOR POWER ELECTRONICS

INTRODUCTION

In the previous two chapters, the component material and shape properties of components for power electronic systems were reviewed. This chapter will list these components that are available commercially.

6.1- TDK FERRITE POWER ELECTRONIC COMPONENTS

TDK offers a wide variety of materials and component shapes for power electronic applications. The materials will be considered first followed by the cores that are offered in the various materials. The specifications listed for each core as it applies to design will be reviewed. The catalog pages for material and representative listings of core properties are found in the Appendix at the end of the chapter.

6.1.1-TDK Power Ferrite Materials

TDK has 6 ferrite materials that can be used for power supply transformers and chokes. They are;

1.-PC40- an standard low frequency material in various E-type cores, (ETD, EC, EI, EE, EF, EP, and RM)
2.-PC44- a lower loss improved material in LP, PQ, EPC (low pro file) cores.
3.-PC50- a higher frequency material in PQ,EPC,EP, and RM cores

Tables of magnetic properties are found in Table 6A1. These three materials have a minimum core loss temperature of about 100° C. for continuous service. Recently, TDK has also introduced 2 other materials that have a minimum core loss temperature of 40-80 ° C. for small portable power supplies that operate intermittently and one that has a higher permeability. Another lower temperature material has a moderately high permeability.

4. PC45-has a minimum core loss temperature between 40-50 ° C.
5. PC46-has a minimum core loss temperature between 60-80 ° C.

6. DN50-has a core loss minimum at 50 ° C. and has a higher permeability than the others. It is available in EPC, ER (high power), EEM cores and also in a special EER core whose design gives 13-20% reduced loss.

The properties of the above three materials are found in Table 6A2. Figure 6.1 shows the core loss vs T for several TDK power materials.

For large high power reactor and transformer cores, TDK has, in addition to the previously cited PC40, another material PC22 with a slightly higher saturation value. The high power cores are available in T (Toroids), UU, EC, EIC, PQ, E, EI, PT and SP cores.

In addition to the material specifications, TDK lists some design data for each core such as the effective parameters (A_e, l_e and V_e). For ungapped cores, they list the calculated output power in Watts for 100 KHz.and 500 KHz. (PC50). For gapped cores for chokes, the list the A_L at 1 KHz.,0.5 mA and 100 T (1,000 Gausses).PQ cores have an NI limit for gapped cores and A_L vs gap. Listed is also the temperature rise versus total loss an also core loss at 100 ° C at 100 KHz. and 200 mT (2,000 Gausses).

TDK Common-Mode Choke materials-TDK has three materials for common-mode choke applications. They are HS52,HS72 and HS10. Their permeabilities are respectively 5500, 7500 and 10,000. They are available in T(Toroids), FT, FTR, ET and UU cores. Except for the U-U, they are all continuous-path (no mating surface) cores. Provided are the A_L values and the effective parameters. Toroids are also available in the H5C2 high permeability material. TDK also offers Common-mode filters equivalent circuit model for Spice. See Table 6A3.

TDK EMI Suppressor Materials-TDK lists 6 materials for EMI suppressor cores. They are HF30, HF40, HF50, HF55, HF60 and HF70. Their properties are given in Table 6A4 . From their resistivities, all except HF 60 are NiZn materials, while HS 60 is a MnZn material. These materials are available in multi-hole substrates, chip suppressors, beads, wire-wound-beads, cable clamps, and toroids. TDK also makes their own EMI filters.

6.2- PHILIPS (YAGEO) POWER FERRITE COMPONENTS

Philips has 12 ferrite materials for power applications. Eleven of them are MnZn ferrites and one NiZn ferrite. The properties of these materials are listed in Table on page . The first 3 materials are used for line output transformers for TV deflection yokes. The remaining 9 materials are arranged essentially in order of their frequency of operation. 3C81 has a minimum core loss minimum at 50 ° C. 3C94 is an industrial use material. For 400 KHz.

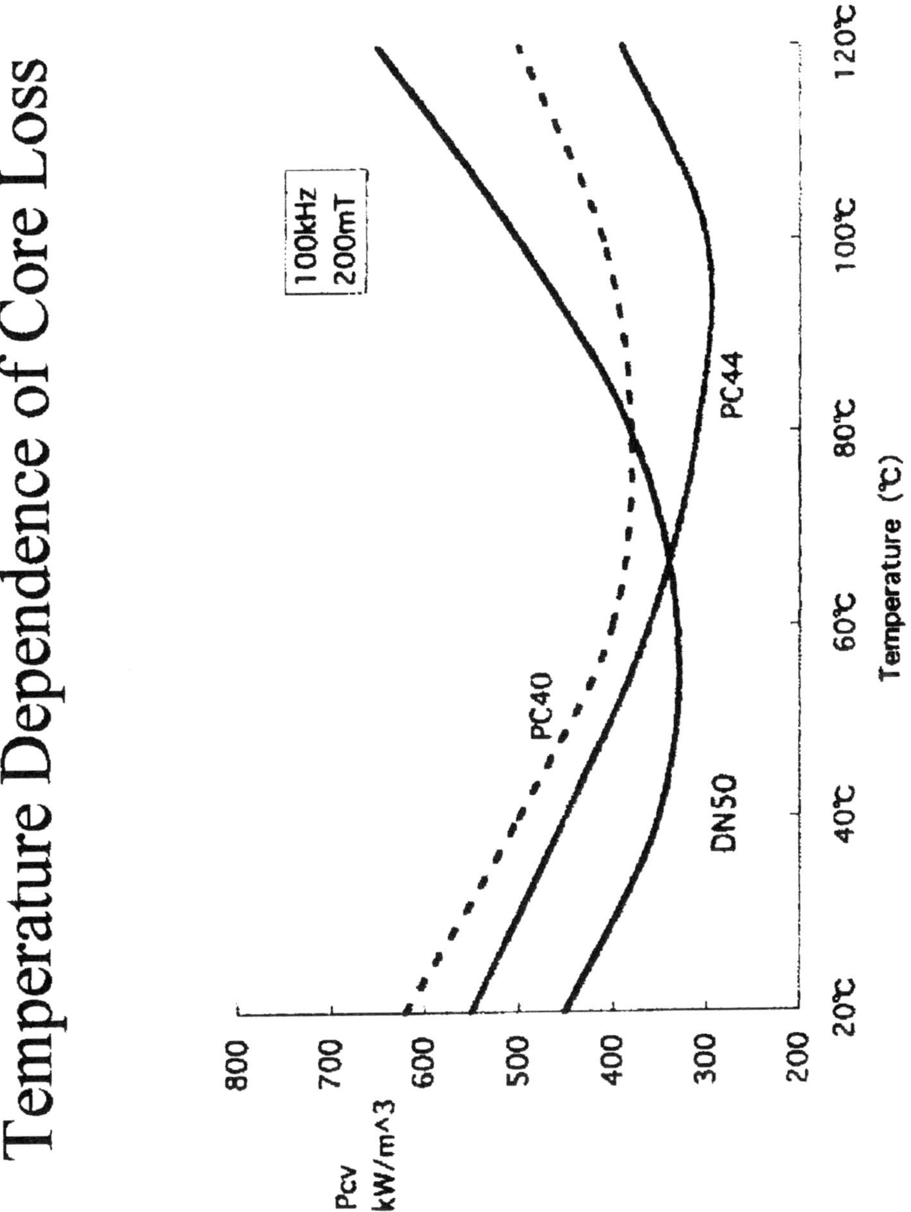

Figure 6.1-Watt Loss vs Temperature for several TDK Power materials

operation, 3C94 is a low loss high B_m material and 3C9 a very low loss material. The 3F materials are for higher frequencies, 3F3 for 700 KHz, 3F35 for 1MHz. and 3F4 for 3 MHz. A graph of the performance factor(PF) defined earlier in Section 3.14 is given in Figure 6A6 for a 500 mW/cm^3 limit for for 5 representative material grades. For selecting a core according to power output, Table 6.1 lists the core sizes capable of handling the power ranges at 100 KHz.

Table 6.1-Power Handling Capacities of Ferrite Cores-Philips

Power throughput for different core types (at 100 kHz switching frequency)

POWER RANGE (W)	CORE TYPE
<5	RM4; P11/7; R14; EF12.6; U10
5 to 10	RM5; P14/8
10 to 20	RM6; E20; P18/11; R23; U15; EFD15
20 to 50	RM8; P22/13; U20; RM10; ETD29; E25; R26/10; EFD20
50 to 100	ETD29; ETD34; EC35; EC41; RM12; P30/19; R26/20; EFD25
100 to 200	ETD34; ETD39; ETD44; EC41; EC52; RM14; P36/22; E30; R56; U25; U30; E42; EFD30
200 to 500	ETD44; ETD49; E55; EC52; E42; P42/29; U37
<500	E65; EC70; U93; U100

The individual core sheets list the A_L ,μ_e and B_{sat} at H = 250 A/m and T = 100 ° C. as well as the core loss at 25, 100 and 400 KHz. Gappe core data lists the A_L ,μ_e and the air gap. In addition , for RM cores, the core loss is given at 1MHz. and 3 MHz. and planar E-cores at 30 and 10 mT respectively.

Philips Common-Mode Choke materials-Philips offers 5 high perm materials that can be used for common-mode choke applications at lower frequencies. They are 3E25, 3e27, 3E26, 3E5 and 3E6 which range in permeabilities respectively from 6000, 6,000, 7,000, 10,000 and 12000.They are mostly available in beads, toroids and other common mode shapes. For higher frequencies SMD common-mode chokes are available in the NiZn material, 4S2.

Philips EMI Suppressor Materials-Philips lists a variety of EMI suppression materials. They are listed in Table 6A15 . The 3E series are MnZn common-mode choke materials. 3S1 is also a rather high perm material that can not withstand a high DC bias and instead the lower perm 4S2 should be used. Two new materials,3S3 and 3S4 are MnZn with high resistivity. All of the remaining materials (4 series)are NiZn materials with high resistivities and lower perms. An impedance versus frequency plot for 3S4 is shown in Figure 6A5.

6.3- EPCOS POWER FERRITE COMPONENTS

Epcos (formerly Siemens-Matsushita) has 10 different power materials. For 100 KHz., there are N27,N53 and N41; for 200 KHz. there are N62,N67,,N72, and N82; for 500 KHz. there is N87; for 300 Khz. to 1MHz. and resonance converters, there are N49 and N59. See Tables 6A17-19 anf Figure 6A6Just released at this book's publication is a new material N97 which listed best in class at 25-400 KHz. The loss at 100 ° C. is 20% lower than that of N87. It also claims a better DC bias property than N87. They also list a new material, N92 with which they claim it is possible to increase the rated current of output chokes by 10% against N87.At the same time, the losses at 100° C. are comparable to N87. They also claim that their N49 material has been improved to meet the needs of rising performance requiements of DC to DC converters. The aim was to reduce losses while increasing saturation. At 500 KHz. the losses have been reduced by 30%. and saturation has been increased by 10%. The material data and The material data and core loss curves for all three materials are given in Tables 6.2-6.4 and Figures 6.2, 6.3, 6.5, 6.7 and 6.8 . In addition DC bias curves are given for N92 and N97 in Figures 6.4 and 6.6 .

A table of transformer power capabilities at several frequencies is given for each core. Epcos lists performance factor, PF for 100 ° C. and 300 kW/m^3 and a rather little cited property, the standardized hysteresis constant as a function of temperature. There is a very wide variety of core shapes. The RM cores are particularly featured in sizes from RM4 to RM 14. They are also available in low profile in this range. Other cores include EP, standard pot cores, E-cores including ELP and EFD. Some cores including the RM's are available with surface -mount accessories.

Individual sheets for cores include A_L ,μ_e and core loss at 100 KHz. and 500 KHz. and 100 ° C. at 200 mT and 50 mT. respectively. Gapped core data show the A_L ,μ_e and the air gap. Low profile and planar cores are available in a several sizes.

Epcos Common-Mode Choke materials-Epcos does not specifically list common-mode choke materials but they have 5 materials that have permeabilities of 6000 and higher (up to 15,000). They are T35,T37,T38 T42 and T46. The last one is only available in small ring cores.

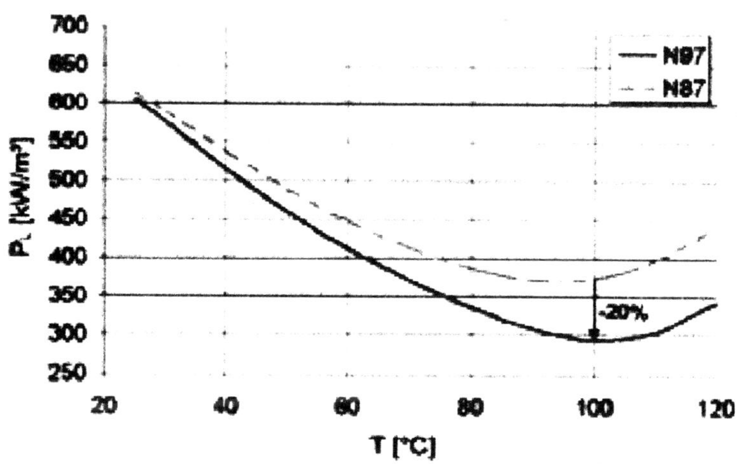

Figure 6.2 . Core loss curves for new material N97 compared with N87

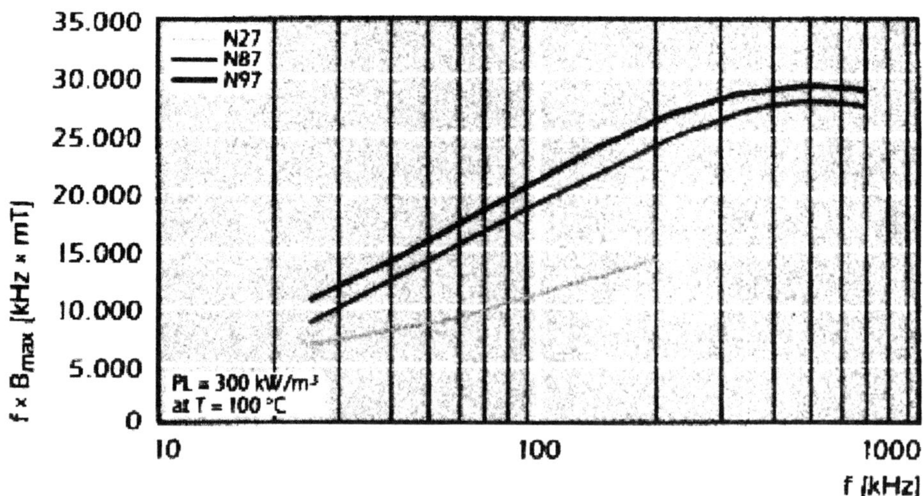

Figure 6.3.- Performance factor of N97 compared with N87

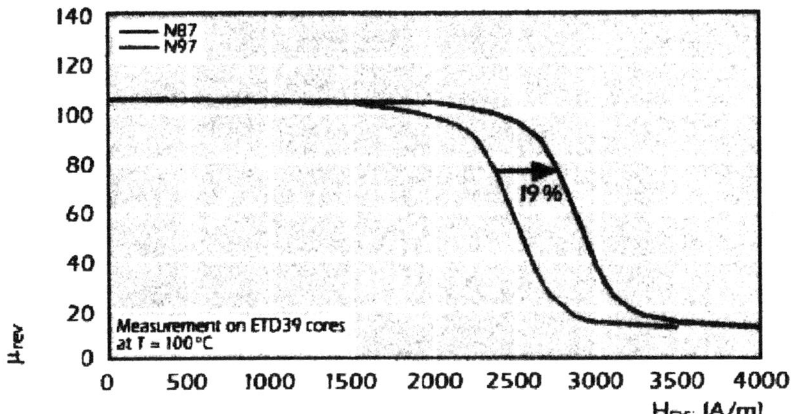

Figure 6. 4- . Reversible Permeability of N97 compared with N87

Table 6.2 -Material Data for N97

Material data

Initial permeability	25°C	μ_i		2300 ± 25%
Flux density, dyn.	25°C	B_{max}	[mT]	510
10 kHz, 1200 A/m	100°C	B_{max}	[mT]	410
Coercive field strength	25°C	H_C	[A/m]	21
f = 10 kHz	100°C	H_C	[A/m]	12
Curie temperature		T_C	[°C]	>230
Relative core losses		P_V		
	100°C	25 kHz 200 mT	[kW/m³]	45
	100°C	100 kHz 200 mT	[kW/m³]	300
	100°C	300 kHz 100 mT	[kW/m³]	340
	100°C	500 kHz 50 mT	[kW/m³]	205
Resistivity		ρ	[Ωm]	8
Density			[kg/m³]	4920

Table 6.3 Material Data for N92

Material data

Initial permeability	25°C	μ_i		1500 ± 25%
Flux density, dyn.	25°C	B_{max}	[mT]	500
10 kHz, 1200 A/m	100°C	B_{max}	[mT]	440
Coercive field strength	25°C	H_C	[A/m]	24
f = 10 kHz	100°C	H_C	[A/m]	13
Curie temperature		T_C	[°C]	>280
Relative core losses		P_V		
	100°C	25 kHz 200 mT	[kW/m³]	70
	100°C	100 kHz 200 mT	[kW/m³]	410
	100°C	300 kHz 100 mT	[kW/m³]	410
	100°C	500 kHz 50 mT	[kW/m³]	230
Resistivity		ρ	[Ωm]	8
Density			[kg/m³]	4850

Core loss vs. temperature

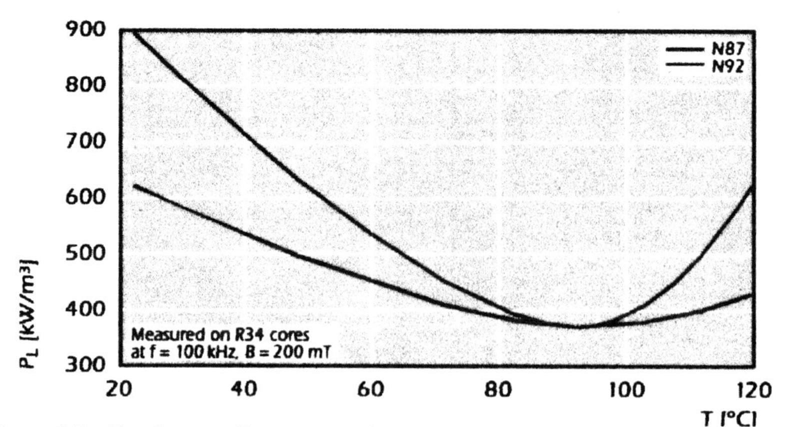

Figure 6.5- Core Loss vs Temperature for N92

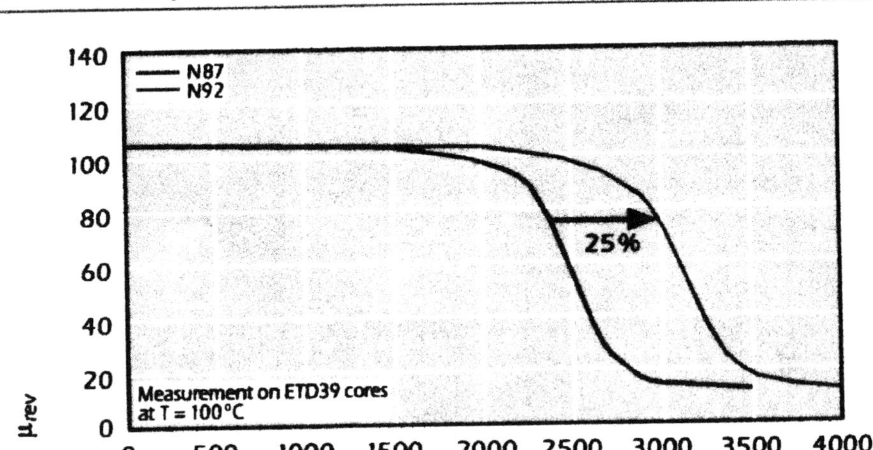

Figure 6.6- Reversible Permeability vs DC Bias for N92 compared to N87

6.4-MAGNETICS POWER FERRITE COMPONENTS

Magnetics has 4 different power materials. One material, F, has a minimum core loss temperature at about 25 ° C and is meant for low frequencies. K and P materials have core loss minimum temperatures at about 60 ° C at 100 and 500 KHz. R has the lowest core loss at 100 ° C for 100 KHz and 500 KHz. K material has the lowest core loss at 700 KHz. See Table 6A5

The various materials are available in all the popular core shapes. Included in the power ferrite design information is a table of core cross section-window area products for all the power cores. Graphs showing the same area product plotted against the output power allowing the core selection to be made for the various core shapes and sizes. Another table gives the power handling capabilities for different frequencies and core shapes for a forward converter. Output power is also plotted against temperature rise for different frequencies. The individual core size sheets list the core-window product area and the A_L for ungapped cores .

Using the common core-loss equation;

$$P_L = af^cB^d$$

Where P_L = the core loss in mW/cm^3

Table 6.4 Material Data for Improved N49

Initial permeability	25°C	μ_i		1500 ± 25%
Flux density, dyn.	25°C	B_{max}	[mT]	490
10 kHz, 1200 A/m	100°C	B_{max}	[mT]	400
Coercive field strength	25°C	H_c	[A/m]	38
f = 10 kHz	100°C	H_c	[A/m]	33
Curie temperature		T_c	[°C]	>240
Relative core losses		P_V		
	100°C	300 kHz 100 mT	[kW/m³]	330
	100°C	500 kHz 50 mT	[kW/m³]	80
	100°C	1000 kHz 50 mT	[kW/m³]	475
Resistivity		ρ	[Ωm]	17
Density			[kg/m³]	4800

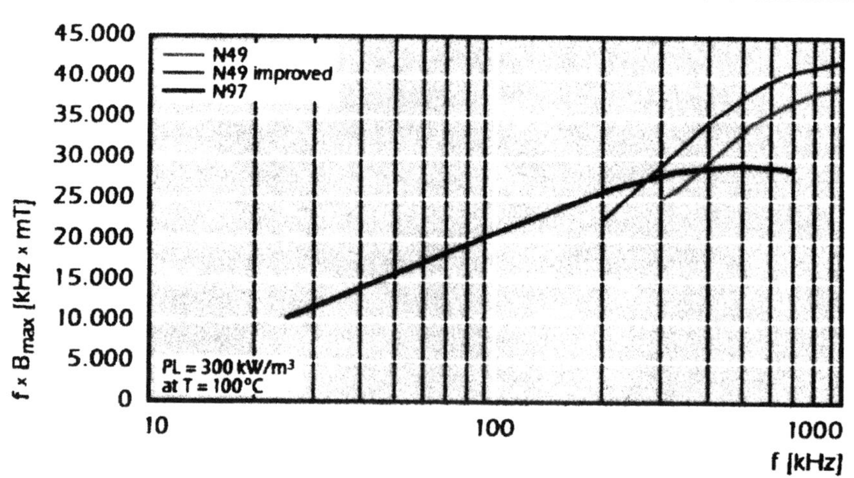

Figure 6.7- Performance Factor of Improved N49 compared to old N49 and new N97

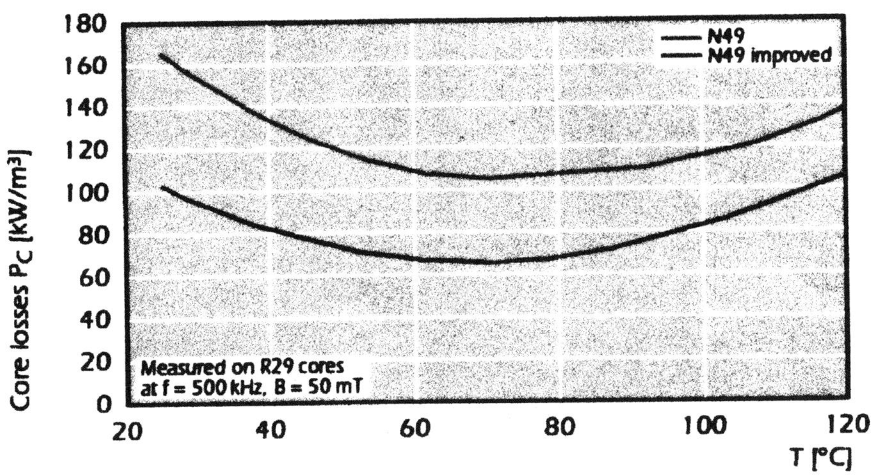

Figure 6.8- Core Loss Curve of Improved N49 compared with that of old N49

f = frequency, Hz.
B = Induction in KG.

Table 6A6 gives the values of the 3 factors for the different materials and frequencies.

Magnetics Common-Mode Choke materials- Magnetics has 3 high permeability materials for common mode choke operation. They are J, W and H materials whose permeabilities are 5,000, 10,000 and 15000 respectively. The latter is available only in toroids. The others can be had in most other core shapes. See Table 6A7

6.5 TOKIN POWER FERRITE COMPONENTS

Tokin's power ferrite line is listed in Table 6A10 .Four materials are listed spanning the range from 100 KHz. to 1 MHz. The core loss minimum temperature for the BH1 and BH2 are in the range 90-100 ° C while that of BH40, the higher frequency material, is between 70-80 ° C. Properties of new material with somewhat higher saturation, BH3 is listed in Table 6A11 compared with BH1. Plots of core loss versus frequency and flux density at three temperatures are presented.

Tokin Common-Mode Choke materials- Tokin has one material, BH5000 which has a higher permeability(5000) for lower frequency common-mode choke application.

6.6- FAIR-RITE POWER FERRITE COMPONENTS

Fair-Rite Products has two materials for use as power transformers, 77 material for 25 Khz. operation and 78 with a saturation of 5000 Gausses for 100 KHz. operation. The 75 material can be used at 25Khz. at 100 ° C. Power materials are available in EP, PQ, U, E and ETD cores. See Table 6A8

Fair-Rite Common-Mode Choke materials- Fair-Rite lists one material 75 for their common-mode choke aplications. It has a permeability of 5000.Another material, 76 has a 10000 permeability and is recommended for frequencies up to 500 Khz.

Fair-Rite EMI Suppressor Materials -Fair-Rite lists 5 materials for EMI suppression, 43, 44, 61,73 and 77. The 73 and 77 materials are MnZn for lower frequencies(<30 MHz.) while the other 3 are NiZn for higher frequencies (30-250 MHz.). The 61 material is the most stable with temperature. See Table 6A9.

6.7-FERRONICS POWER FERRITE COMPONENTS

Ferronics Common-Mode Choke materials-Ferronics has 2 MnZn ferrite materials, B (5000) and T(10,000) perm which can be used for common mode chokes at lower frequencies. See Table 6A23

Ferronics EMI Suppressor Materials- Ferronics has 3 NiZn materials that are used for EMI suppression, J, K, and P. The permeabilities are 850, 125 and 40 respectively. The latter two are perminvar materials that list the caution that the permeabilities and loss factors may be irreversibly increased if excited with a high magnetizing force. This factor should be considered when applying high DC or ac currents. J perm is recommended for frequencies from 5-500 MHz, K for above 20 Mhz. and P for above 80MHz. The materials are offered in toroids, multi-hole wide band cores and beads.The material properties are given in Table 6A23 .

6.8-FDK POWER FERRITE COMPONENTS

FDK lists 7 different power ferrite materials as shown in Table 6A25. The 6H series are high flux density materials for 25-100 KHz. operation. The higher the number after the 6H, the lower the losses. The minimum loss temperatures vary from 40 ° C - 100 ° C. The 7H series are lower flux density materials for higher frequencies. The newest material 7H20 has half of the core loss of 7H10 at 1 Mhz. and 100 ° C. The individual core data include the effective parameter, the window and core cross-sectional areas and the A_L's. See Fig 6A7-8

FDK Common-Mode Choke materials- FDK recommends two high permeability materials, 2H07 and 2H10 with permeabilities of 7,000 and 10,000 respectively for common mode usage. 2H10 is recommended for frequencies lower than 500 KHz.

FDK EMI Suppressor Materials-FDK lists 4 EMI suppressor materials, K32, L51, K14, and K26. They are NiZn materials ranging in permeability from 700 to 40 over a range of increasing frequencies. Their properties are listed in Table 6A24 .

6.9-AVX POWER FERRITE COMPONENTS

The AVX ferrite product line is that of the former Thomson company. Their power ferrite component line is made from 8 materials listed in Table 6A26. The PW1 materials are for the lower frequencies up to 32 KHz. probably for line output TV transformers. The PW2 materials operate at 100 KHz. The PW3 and PW4 materials run up to 500KHz. and the PW5 to 1.5 MHz. See Table 6A26.

AVX Common-Mode Choke materials-There are 3 high perm AVX materials that can serve as common mode chokes. They are A2, A3 and A4 with permeabilities of 10,000, 7500 and 6000 respectively. Their properties are given in Table 6A27 .

6.10-MMG-NEOSID POWER FERRITE MATERIALS

MMG has six materials for power transformers and chokes, F47, F45, F44, F5, F5A and F5C. Their permeabilities range from 1800 to 3000. Core loss data for 25 ° C. and 100 ° C. is shown for the F5 series at 16 and 25 KHz., F44 and F45 up to 100 KHz. and F47 to 400 KHz. See Table 6A12.

MMG-Neosid Common-Mode Choke materials- MMG has 4 materials with permeabilities of 5000 and above. They are the F9C, F10 and F39 whose permeabilities are 5000, 6000, and 10,000.

MMG-Neosid EMI Suppressor Materials-MMG-Neosid list 4 materials for EMI suppression, F19, F14, F16, F25, F28 and F29, the latter three being Perminvar ferrites for higher frequencies. Their permeabilities are 1000, 220, 125, 50, 30 and 12 respectively.

6.11-KASCHKE POWER FERRITE MATERIALS

Kaschke has two materials for use as power transformers' K2004 and K2006. The K2004 has a higher saturation and lower losses. The losses are given at 16 KHz. The 2004 material has a minimum loss temperature of 80 ° C while that of 2006 at near 100 ° C. See Tables 6A28-29 and Figure 6A9.

Kaschke Common-Mode Choke materials- Kaschke has one material K6000 with a permeability of 6000 that may be used for common-mode chokes

6.12- VOGT POWER FERRITE MATERIALS

Vogt has three power ferrite materials, Fi323,Fi324 and Fi325. All three have saturations in the 5000 Gauss range with the 325 material slightly higher than the other two it also has the lowest losses at the higher frequencies. The other two are meant for the 100-300 KHz range with the 324 material having a higher minimum loss temperature than the 323 (80-100 ° C. vs 60 ° C. See Table 6A31.

Vogt Common-Mode Choke materials-Vogt has three materials suitable for common-mode chokes. They are Fi410, Fi360 and Fi 350 with permeabilities of 10,000, 6000 and 5,000. Their properties are listed in Table 6A31.

6.13-SAMWHA POWER FERRITE MATERIALS

Samwha has three power ferrite materials, PL-5, PL7 and PL9. The saturations for all three are about the same at 5000 Gausses. The core loss at 100 KHz and 100 ° C is lowest for the PL9 then the PL7 and then the PL5. The minimum loss temperature for the PL9 is about 80 ° C while those for the other two are about 95 ° C. Samwha also lists three other materials for flyback transformer operation, SM19B, SM 19C and SM19D. All the magnetic parameters are identical to the other three power materials except that the core loss is quoted at 32 KHz. See Table 6A39.

Samwha EMI Suppressor Materials-Samwha lists 4 NiZn materials for EMI suppression, SN20, T314, SN065 and SN201. They range down in permeability from 2000 to 500.

6.14- STEWARD POWER FERRITE MATERIALS

Steward does not specifically list a power transformer material but does recommend their 21(presumably a NiZn Material) material for temperature-stable high frequency inductor and some very high frequency choke materials. They list several EMI suppression materials including their 25 and 38 materials. Their materials are listed in Table

Steward Common-Mode Choke materials-Steward recommends their 1700 perm 38 material for broadband common-mode choke applications. They do however have some high perm materials for the lower frequencies. The are; 35 and 36 (5000 perm), 37 and 42 (7500 perm) and 40 (10,000 perm). All of these are listed in their toroid catalog. (see Tables 6A20-1.)

Steward EMI Suppressor Materials- Steward lists 3 EMI suppresion materials, 25, 28 and 29. Their permeabilities range from 125 to 850. They are nickel ferrites with high resistivities. Steward makes a very long line of EMI cores and filters. Their material specs are given in Table 6A22 .

6.15-FERRITE INTERNATIONAL POWER FERRITES

Ferrite International has 4 materials for power ferrite applications, TSF 5099 TSF7099, TSF7070 and TSF8040. The first three have saturations of 5000 Gausses with TSF8040 at 5100 for integrated magnetics. TSF5099 has the lowest core loss at 25 and 100 KHz. and 100 ° C. TSF 5099 and 7099 have core has minimum temperatures at about 100 ° C, TSF7070 at 80 ° C and 8040 at 60 ° C. They also list a "Boost" material for DC bias operation in gapped inductors. Their material properties are shown in Table 6A37 .

Ferrite International Common-Mode Choke materials-Ferrite International has two high perm materials for common-mode choke usage. They are TSF5000(5000 perm and TSF010K (10,000 perm). See Table 6A37.

6.16-CERAMIC MAGNETICS POWER FERRITE MATERIALS

Ceramic Magnetics has 4 power ferrite materials MN80, MN67, MN60LL and MN8CX. The first three have data at 125 ° C. MN80 shows core loss data at 3000 Gausses and 100 KHZ. with a minimum core loss temperature of 75 ° C. Under the same frequency at 1000 Gausses, the minimum core loss temperature is 125 ° C. MN60LL has the lowest core losses at 200-300 KHz. MN67 has the highest saturation at 5200 Gausses. MN8CX has the highest operating frequency at .5-2.0 MHz. See Table 6A38.

Ceramic Magnetics Common-Mode Choke materials-Ceramic Magnetics lists 4 materials wth permeabilities above 5000. They are MN100 (10,000 perm ,MC25 (9500 perm) and MN60 and MN60LL (6000 perm)

6.17-TOMITA POWER FERRITE MATERIALS

Tomita lists two power ferrite materials, 5G and 15 G. Data is given at 16 KHz. and 25 ° C and 100 ° C. also at .1 and .2 T up to 100 KHz. Core shapes include RM, X, U, E, EP ETD, PM and toroids. See Table 6A30.

Tomita Common-Mode Choke materials-Tomita has 5 materials whose permeabilities are between 5,000 and 10,000. They are 2E2, 2E2B, @#1, 2G1, 2G3 and 2F1.

6.18-ISKRA POWER FERRITE MATERIALS

Iskra lists core loss data for three materials although others are shown with high saturations but no loss data. The materials with core loss data are; 2E6 2F6 and 2F8 at 100, 200 and 300 KHz. respectively. Core shapes listed for these materials are; EE, EI, EP, RM, EER, ETD, EC and PP.See Table 6A33.

Iskra Common-Mode Choke materials-Iskra lists 2 materials with permeabilities above 5000. They are 22G (6,000 perm and 12G (10,000 perm)

6.19- DOMEN POWER FERRITE MATERIALS

Domen lists two materials for High power applications, 2500HMC1 and 2500HMC2. The losses are given at 16 KHz. 2000 Gausses at 25 ° C and 100 ° C. See Table 6A32

Domen Common-Mode Choke materials-Domen has 3 materials with high permeability , 6000HM and 6000HM1 at 6000 perm and 10000HM at 10,000 perm.

6.20- HITACHI POWER FERRITE MATERIALS

There are 4 power ferrite materials in the Hitachi catalog, SB5S, SB3L, SB7C, and SB9C. SB5S has a higher permeability (3000) than the others but higher losses at 100 KHz. and 2000 Gausses. SB7C and SB9C have low losses at higher frequencies with a core loss maximum at 100 ° C. SB3L has a high saturation and is particularly useful with imposed DC bias. These materials are available in EI, EE, EER, and PQ shapes.See Table 6A35.

Hitachi Common-Mode Choke materials-Hitachi lists several materials for common-mode chokes. They are GP11, GP9,GP7 GP5 and GQ5C with permeabilities from 5,000 to 10,000 for use at lower frequencies.

Hitachi EMI Suppressor Materials-Hitachi lists 4 NiZn materials (DL) for use in EMI suppression up to 10 MHZ.(See Table 6A36) At 100 MHz. materials QM, KP, DV and SH are recommended but no data given. Shapes available are beads or filters.

6.21- COSMO POWER FERRITE MATERIALS

Cosmo has four power ferrites listed in their catalog. They are CF138, CF129, CF196, and CF101. The power losses are given at 16 KHz and 25 KHz at a 2000 Gauss level.The saturation of CF129 is 5100 Gausses and the other range from4800 to 5000 Gausses. An unusual parameter they list is the temperature of the secondary permeability maximum, which should correspond to the minimum core loss temperature. The materials are available in EFD, EPC,EE, EI, ETD, EER, EC and UU. shapes.

Cosmo Common-Mode Choke materials- Cosmo has two materials CF195 and CF197 with permeabilities of 5000 and 7500. See Table 6A40.

6.22- ACME POWER FERRITE MATERIALS

Acme lists 3 power ferrite materials, P2, P4 and P5. The saturation of P2 is 4500 Gausses and the other two are at 4800 Gausses. P2 is meant for 100 KHz operation , P4 at 300 KHz. and P5 at 500 KHz. Acme lists an unusual property namely the amplitude permeabilities at 25 KHz. and 2000 Gausses at 25 ° C. and 100 ° C. See Table 6A43

Acme Common-Mode Choke materials- Acme lists several high permeability materials that can be used for common-mode chokes.

Acme EMI Suppressor Materials-Acme-Malaysia lists a number of NiZn materials for EMI suppression in Table 42. They are H2, H3 H4, H5, D1C and D28.Their permeabilities range from 50 to 870. They are available in several shapes including cable cores.

6.23- HINODAY POWER FERRITE MATERIALS

Hinoday was the former Morris Electronics Co. an old Indian ferrite company. Hinoday has 3 ferrite power materials MSB-58, MSB-7C and MSP-5F. The saturation of the first is 4800 Gausses while the other 2 are at 5000 Gausses. The MSB-7C has core loss listed at 100 KHz. and 2000 Gausses while the first is easured at 16 KHz. and 2000 Gausses at 40 ° C. The last one is also a 16KHz. material with the higher flux density. See Table 6A44

Hinoday Common-Mode Choke materials-Hinoday has 2 materials MGQ5C and MGP-9 with permeabilities of 5300 and 7000 for common-mode chokes.The properties are given in Table 6A44 .

6.24- ISU POWER FERRITE MATERIALS

Isu has 6 power ferrite materials, PM, PM2, PM2A, PM5 PM7 and PM9. The first 3 are for lower frequencies from 16-25 KHz. and 85 ° C. while the other three are mainly for 100 KHz. and 100 ° C. operation. The saturations of the first three are also slightly lower than the last three. See Table 6A41

Isu Common-Mode Choke materials- Isu does not list any materials for common-mode chokes

6.25- MIANYANG POWER FERRITE MATERIALS

Mianyang has 3 power ferrite materials, R2KD, R2KH and R2KBP1. The first has a saturation of 4800 Gausses and the second 5100 Gausses. The first material is meant for 16KHz operation , the second for 16,32 and64 KHz. while the third is a 25 KHz. material. See table 6A45.

Mianyang Common-Mode Choke materials-No common-mode materials listed.

6.26- HEBEI POWER FERRITE MATERIALS

Hebei has one power ferrite material for operation at 25 KHz and 2000 Gausses. The saturation is high at 5100 Gausses. See Table 6A46

Appendix 6.1

Listing of Catalog Data for Ferrite Core Suppliers

Tables Appendix List No.	Vendor	Figures Appendix List No.
1. 6A1	TDK	
2. 6A2	"	
3. 6A3	"	
4. 6A4	"	
5. 6A5	Magnetics	

6. 6A6	"	
7. 6A7	"	
8. 6A8	Fair-Rite	
9. 6A9	"	
10. 6A10	Tokin	
11. 6A11	"	
12. 6A12	MMG-Neosid	
13. 6A13	Philips	6A1,6A2
14. 6A14	"	6A3
15. 6A15	"	6A4
16. 6A16	"	6A5
17. 6A17	Epcos	6A6
18. 6A18	"	
19. 6A19	"	
20. 6A20	Steward	
21. 6A21	"	
22. 6A22	"	
23. 6A23	Ferronics	
24. 6A24	FDK	6A7
25. 6A25	"	6A8
26. 6A26	AVX	
27. 6A27	"	
28. 6A28	Kaschke	6A9
29. 6A29	"	
30. 6A30	Tomita	
31. 6A31	Vogt	
32. 6A32	Domen	
33. 6A33	Vogt	
34. 6A34	Iskra	
35. 6A35	Hitachi	
36. 6A36	"	
37. 6A37	TSC	
38. 6A38	CMI	
39. 6A39	Samwha	
40. 6A40	Cosmo	
41. 6A41	Isu	
42. 6A42	Acme	
43. 6A43	"	
44. 6A44	Hinoday	
45. 6A45	Mianyang	
46 6A46	Hebei	

Table 6A.1-Properties of TDK Power Transformer and Choke Materials

FOR TRANSFORMER AND CHOKE

Material					PC40	PC44	PC50
Initial permeability	μi				2300±25%	2400±25%	1400±25%
Amplitude permeability	μa				3000min.	3000min.	
Core loss [B=200mT]	Pcv	kW/m³	25kHz sine wave	25°C	120		
				60°C	80		
				100°C	70		
				120°C	85		
			100kHz sine wave	25°C	600	600	130[*2]
				60°C	450	400	80[*2]
				100°C	410	300	80[*2]
				120°C	500	380	
Saturation magnetic flux density[*1] [H=1194A/m]	Bs	mT		25°C	510	510	470
				60°C	450	450	440
				100°C	390	390	380
				120°C	350	350	
Remanent flux density[*1]	Br	mT		25°C	95	110	140
				60°C	65	70	110
				100°C	55	60	98
				120°C	50	55	
Coercive force[*1]	Hc	A/m		25°C	14.3	13	36.5
				60°C	10.3	9	31
				100°C	8.8	6.5	27.2
				120°C	8	6	
Curie temperature[*1]	Tc	°C			>215	>215	>240
Electrical resistivity[*1]	ρv	Ω-m			6.5	6.5	
Density[*1]	db	kg/m³			4.8×10^3	4.8×10^3	4.8×10^3

*1 Average value

*2 500kHz, 50mT

• The values were obtained with toroidal cores at room temperature unless otherwise shown.

Table 6A2- Properties of TDK Power Materials

Material				PC45	PC46	PC44 (Conventional material)
Initial permeability	μi			2500±25%	3200±25%	2400±25%
Power loss [100kHz, 200mT]	Pcv	kW/m³		570[25°C] 250[75°C] 460[100°C]	350[25°C] 250[45°C] 660[100°C]	600[25°C] 400[60°C] 300[100°C]
Saturation magnetic flux density	Bs	mT	[25°C] [100°C]	530 420	530 410	510 390
Remanent flux density	Br	mT	[25°C] [100°C]	120 80	115 80	110 60
Coercive force [1194A/m]	Hc	A/m	[25°C] [100°C]	12 8	11 10	13 6.5
Curie temperature	Tc	°C		≧230	≧230	≧215

Table6A3-Properties of TDK Common-Mode Choke Materials

FOR EMC PREVENTION COMMON-MODE CHOKE

Material			HS52	HS72	HS10
Initial permeability	μi		5500±25%	7500±25% (2000min. at 500kHz)	10000±25%
Relative loss factor	$\tan\delta/\mu i \times 10^{-6}$		10 (100kHz)	30 (100kHz)	30 (100kHz)
Saturation magnetic flux density* [H=1194A/m]	Bs	mT	410	410	380
Remanent flux density*	Br	mT	70	80	120
Coercive force*	Hc	A/m	6	6	5
Curie temperature*	Tc	°C	>130	>130	>120
Electrical resistivity*	ρv	Ω-m	1	0.2	0.2
Density*	db	kg/m³	4.9×10^3	4.9×10^3	4.9×10^3

* Average value

• The values were obtained with toroidal cores at room temperature unless otherwise shown.

Table 6A4.Properties of TDK EMI Suppression Materials

Material	μ_{iac}	Temperature coefficient $\alpha\mu r$ ($\times 10^{-6}$/°C)	Bms (mT) at H (A/m)	T_C (°C) [°F] min.	ϱ (Ω-m)
HF70	1500	1 to 3	280 (1600)	100 [212]	10^5
HF60	900	8 to 14	300 (1600)	130 [266]	10^2
HF55	550	15 to 35	290 (1600)	125 [257]	10^5
HF50*	250	9 to 15	290 (1600)	125 [257]	10^5
HF40	120	8 to 18	410 (4000)	250 [482]	10^6
HF30	45	5 to 15	320 (4000)	300 [572]	10^5

Table 6A5-Magnetics Power Materials

POWER MATERIALS SUMMARY

		F	P	R	K	J	W+
μi(20 gauss)	25°C	3000	2500	2300	1500	5000	10,000
μp(2000 gauss)	100°C	4600	6500	6500	3500	5500	12,000
Saturation Flux Density Bm Gauss	25°C	4900	5000	5000	4600	4300	4300
	100°C	3700	3900	3700	3900	2500	2500
Core Loss (mw/cm³) (Typical) @100 kHz, 1000 Gauss	25°C	100	125	140	100		
	60°C	180	80*	100	90		
	100°C	225	125	70	110		

* @80°C + @10kHz

Table 6A6- Parameters in Power Loss Equation-From Magnetics

		a	c	d
K Material	f<500 kHz	0.0530	1.60	3.15
	500 kHz≤f<1 MHz	0.00113	2.19	3.10
	f≥1 MHz	$1.77*10^{-9}$	4.13	2.98
R Material	f<100 kHz	0.074	1.43	2.85
	100 kHz ≤f<500 kHz	0.036	1.64	2.68
	f≥500 kHz	0.014	1.84	2.28
P Material	f<100 kHz	0.158	1.36	2.86
	100kHz≤f<500 kHz	0.0434	1.63	2.62
	f≥500 kHz	$7.36*10^{-7}$	3.47	2.54
F Material	f≤10 kHz	0.790	1.06	2.85
	10 kHz<f<100 kHz	0.0717	1.72	2.66
	100 kHz≤f<500 kHz	0.0573	1.66	2.68
	f≥500 kHz	0.0126	1.88	2.29
J Material	f≤20 kHz	0.245	1.39	2.50
	f>20 kHz	0.00458	2.42	2.50
W Material	f≤20 kHz	0.300	1.26	2.60
	f>20 kHz	0.00382	2.32	2.62
H Material	f≤20 kHz	0.148	1.50	2.25
	f>20 kHz	0.135	1.62	2.15

Table 6A7-Magnetics Common-Mode Choke Materials

EMI/RFI FILTERS AND BROADBAND TRANSFORMERS		
J	W	H
5000 ± 20%	10000 ± 30%	15000 ± 30%
<1	<.25	<.15
<20 (100kHz)	<7 (10kHz)	<15 (10kHz)
>140	>125	>120
4300 430	4300 430	4200 420
1000 100	800 80	800 80
0.1 8	0.04 3	0.04 3
<3	<3	<2.5
1	.15	.1
4.8	4.8	4.9

Table 6A.8 -Fair-Rite Power and Common-Mode Choke Materials

Property	Unit	Symbol	77	78	73	75	76
Initial Permeability @ B <10 gauss		μ_i	2000	2300	2500	5000	10000
Flux Density @ Field Strength	gauss mT oersted A/m	B H	4600 460 10 800	5000 500 37.5 3000	4000 400 10 800	3900 390 10 800	4000 400 2 160
Residual Flux Density	gauss mT	B_r	1150 115	1500 150	1000 100	1250 125	1250 125
Coercive Force	oersted A/m	H_c	.22 17.5	.20 16	.18 14.5	.16 13	.10 8
Loss Factor @ Frequency	10^{-6} MHz	$\tan \delta/\mu_i$	4.5 .1	4.5 .1	7 .1	15 .1	15 .025
Temperature Coefficient of Initial Permeability (20-70 °C)	%/°C		.6	1.2	.8	.9	.9
Curie Temperature	°C	T_c	>200	>200	>160	>140	>120
Resistivity	Ω cm	ρ	10^2	$2\ 10^2$	10^2	$3\ 10^2$	50
Power Loss Density 25kHz - 2000 G - 100°C 100kHz - 1000 G - 100°C	mW/cm³	P	200 –	<115 <130	– –	140 –	– –
Recommended Frequency Range Application Areas L E P S	MHz		<3 <30 <.1 –	<2.5 – <.5 –	<2.5 <30 – –	<.75 <15 <.1 –	<.5 – – –
See this page for additional material data			16 & 17	18 & 19	15	20	21

Table 6A.9 -Fair-Rite EMI Suppression Materials

Property	Unit	Symbol	61	44	43	85*	77	78	73
Initial Permeability @ B <10 gauss		μ_i	125	500	850	900	2000	2300	2500
Flux Density @ Field Strength	gauss mT oersted A/m	B H	2350 235 10 800	3000 300 10 800	2750 275 10 800	3900 390 10 800	4800 460 10 800	5000 500 37.5 3000	4000 400 10 800
Residual Flux Density	gauss mT	B_r	1200 120	1100 110	1200 120	3400 340	1150 115	1500 150	1000 100
Coercive Force	oersted A/m	H_c	1.6 128	.35 28	.30 24	.50 40	.22 17.5	.20 16	.18 14.5
Loss Factor @ Frequency	10^{-6} MHz	$\tan\delta/\mu_i$	32 2.5	85 1.0	120 1.0	50 .1	4.5 .1	4.5 .1	7 .1
Temperature Coefficient of Initial Permeability (20-70 °C)	%/°C		.15	–	1.0	–	.6	1.2	.8
Curie Temperature	°C	T_c	>350	>160	>130	>200	>200	>200	>160
Resistivity	Ω cm	ρ	10^8	10^9	10^5	$2\ 10^2$	10^2	$2\ 10^2$	10^2
Power Loss Density 25kHz - 2000 G - 100°C 100kHz - 1000 G - 100°C	mW/cm³	P	– –	– –	– –	– –	200 –	<115 <130	– –
Recommended Frequency Range Application Areas L E P S	MHz		<100 >200 – –	– 30-250 – –	<10 30-200 – –	– – – <.15	<3 <30 <.1 –	<2.5 – <.5 –	<2.5 <30 – –
See this page for additional material data			10	12	14	11	16 & 17	18 & 19	15

Table 6A.10- Tokin Power Ferrite Materials

Material Characteristics				Unit	BH1	BH2	B40	5000B
Applied frequency range				MHz	<0.3	<0.3	0.3–1.0	
Initial permeability		μ_i			2300±20%	2300±20%	1500±20%	5000±20%
Effective saturation magnetic flux density (approx. 1200A/m)		Bms	23°C 100°C	mT	520 410	510 400	530 440	500 360
Effective retentivity		Brms	23°C 100°C	mT	100 56	100 55	250 120	120 90
Effective coercivity		Hcms	23°C 100°C	A/m	13.0 5.0	14.3 5.0	43.6 27	11.0 6.5
Relative loss factor (100kHz)		$\tan\delta/\mu_i$		10^{-6}	<5	<6	<3	<20
Curie temperature		T_c		°C	220	220	240	180
Core loss	100kHz 200mT	P_{cv}	23°C	kW/m³	550	600		650
			60°C		350	450		500
			100°C		250	410		800
	500kHz 100mT	P_{cv}	60°C				580	
			80°C				500	
	1MHz 50mT	P_{cv}	60°C				360	
			80°C				380	
Density		d		kg/m³	4.8×10^3	4.8×10^3	4.9×10^3	4.8×10^3

Table 6A11-Properties of new Tokin BH3 vs those of BH1

Item				BH-3	BH1
Initial permieability	μi			1800±20%	2500±20%
Effective saturation magnetic flux density	Bms	mT	23°C	530	520
			100°C	420	410
Effective retentivity	Br	mT	23°C	120	100
			100°C	70	55
Effective saturation coercive force	Hc	A/m	23°C	17	13
			100°C	13	5
Core loss	Pcv	kW/m³	25°C	580	550
			60°C	400	350
			100°C	380	250
			120°C	420	300
Cure point	Tc	°C	100°C	260	220

Table 6A12.-MMG-Neosid Power Ferrite Materials

MATERIALS FOR POWER/SWITCHING TRANSFORMERS, DIFFERENTIAL MODE CHOKES, AND OUTPUT CHOKES.

MATERIAL	-	-	F47	F45	F44	F5	F5A	F5C
ORDERING CODE	-	-	47	45	44	25	49	S49
-	-	-	-	-	-	-	-	-
-	STANDARD TEST CONDITIONS	UNITS	-	-	-	-	-	-
INITIAL PERMEABILITY (NOMINAL)	B<0.1mT. 10KHz @ 25C	-	1800	2000	1900	2000	2500	3000
SATURATION FLUX DENSITY (TYPICAL)	H=796 A/m =10 Oe @ 25C	mT	470	500	500	470	470	460
REMANENT FLUX DENSITY (TYPICAL)	H->0 10KHz @ 25C	mT	130	165	270	200	150	150
COERCIVITY (TYPICAL)	B->0...10KHz @ 25C	A/m	24	15	27	21	15	18
CURIE TEMPERATURE (MINIMUM)	B<0.1mT 10KHz	°C	200	230	230	200	200	180
RESISTIVITY (TYPICAL)	1V/cm 25C	ohm/cm	100	100	100	100	100	100
AMPLITUDE PERMEABILITY (MINIMUM)	400mT 25C	-	2000	2500	2500	2400	2400	2400
	320mT 100C		2500	-	-	1825	1825	-
	340mT 100C		-	2000	1900	-	-	-
TOTAL POWER LOSS DENSITY (MAX.)	200mT; 16KHz; 25C	mW/cc	-	-	-	120	120	120
	200mT; 16KHz; 60C		-	-	-	110	110	120
	200mT; 16KHz;100C		-	-	-	110	110	110
	200mT; 25KHz; 25C		120	-	200	-	-	-
	200mT; 25KHz; 60C		-	-	-	190	190	190
	200mT; 25KHz;100C		100	-	130	190	190	190
	100mT; 100KHz; 25C		110	-	250	-	-	-
	100mT; 100KHz; 100C		80	80	160	-	-	-
	200mT; 100KHz; 100C		-	400	750	-	-	-
	50mT; 400KHz; 25C		150	-	-	-	-	-
	50mT; 400KHz; 100C		150	-	-	-	-	-

Table 6A13.-Philips Power Ferrite Materials

Ferrite material	μ_i at 25 °c	B_{sat} (mT) at 25 °c (3000 A/m)	T_C (° C)	(m)	Ferrite type	Main application area
3C15	1800	≈ 500	≥ 190	≈ 1	MnZn	power conversion general purpose transformers
3C30	2100	≈ 500	≥ 240	≈ 2	MnZn	
3C34	2100	≈ 500	≥ 240	≈ 5	MnZn	
3C81	2700	≈ 450	≥ 210	≈ 1	MnZn	
3C90	2300	≈ 450	≥ 220	≈ 5	MnZn	
3C91	3000	≈ 450	≥ 220	≈ 5	MnZn	
3C94	2300	≈ 450	≥ 220	≈ 5	MnZn	
3C96	2000	≈ 500	≥ 240	≈ 5	MnZn	
3F3	2000	≈ 450	≥ 200	≈ 2	MnZn	
3F4	900	≈ 450	≥ 220	≈ 10	MnZn	
3F35	1400	≈ 500	≥ 240	≈ 10	MnZn	
4F1	80	≈ 350	≥ 260	≈ 10^5	NiZn	

Table 6A14-Philips Magnetic Regulator Materials

▮ Ferrites for magnetic regulators

Ferrite material	μ_i at 25 °c	B_{sat} (mT) at 25 °c (3000 A/m)	T_C (° C)	(m)	Ferrite type	Main application area
3R1	≈ 800	≈ 450	≥ 230	≈ 10^3	MnZn	magnetic regulators

Table 6A15-Philips EMI Materials

Ferrite material	μ_i at 25 °c	B_{sat} (mT) at 25 °c (3000 A/m)	T_C (° C)	(m)	Ferrite type	Main application area
3B1	900	≈ 700	≥ 150	≈ 0.2	MnZn	EMI-suppression, tuning
3C11	4300	≈ 340	≥ 125	≈ 1	MnZn	
3E5	10000	≈ 380	≥ 125	≈ 0.5	MnZn	
3E6	12000	≈ 380	≥ 130	≈ 0.1	MnZn	
3E25	6000	≈ 380	≥ 125	≈ 0.5	MnZn	
3E26	7000	≈ 450	≥ 155	≈ 0.5	MnZn	
3E27	6000	≈ 400	≥ 150	≈ 0.5	MnZn	
3S1	4000	≈ 400	≥ 125	≈ 1	MnZn	
3S3	350	≈ 350	≥ 225	≈ 10^4	MnZn	
3S4	1700	≈ 350	≥ 110	≈ 10^3	MnZn	
4A11	700	≈ 320	≥ 125	≈ 10^5	NiZn	
4A15	1200	≈ 340	≥ 125	≈ 10^5	NiZn	
4B1	250	≈ 350	≥ 250	≈ 10^5	NiZn	
4C65	125	≈ 380	≥ 350	≈ 10^5	NiZn	
4D2	60	≈ 240	≥ 400	≈ 10^5	NiZn	
4E1	15	≈ 200	≥ 500	≈ 10^5	NiZn	
4S2	700	≈ 350	≥ 125	≈ 10^5	NiZn	
4S4	250	≈ 300	≥ 130	≈ 10^5	NiZn	
4S7	200	≈ 300	≥ 140	≈ 10^5	NiZn	
4S7	200	≈ 300	≥ 140	≈ 10^5	NiZn	

Table 6A16-Philips 4F1 Material

SYMBOL	CONDITIONS	VALUE	UNIT
μ_i	25 °C; ≤10 kHz; 0.1 mT	≈80	
μ_a	100 °C; 25 kHz; 200 mT	≈300	
B	25 °C; 10 kHz; 250 A/m 100 °C; 10 kHz; 250 A/m	≥50 ≥100	mT
P_V	100 °C; 3 MHz; 10 mT 100 °C; 10 MHz; 5 mT	≤200 ≤200	kW/m³
ρ	DC; 25 °C	≈10⁵	Ωm
T_C		≥260	°C
density		≈4600	kg/m³

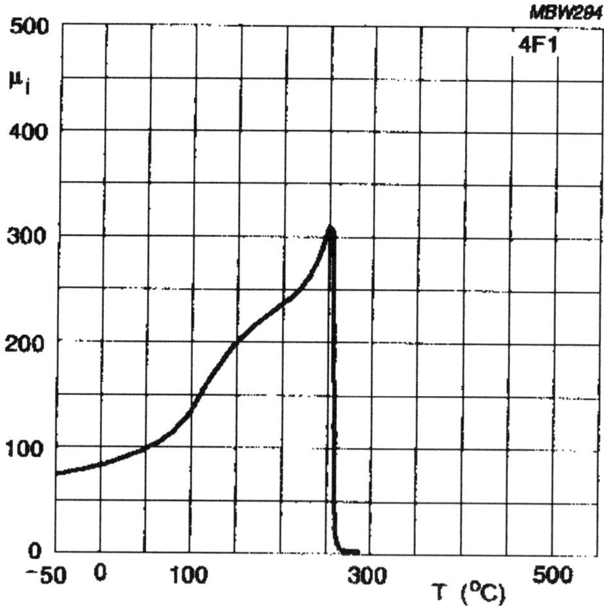

Figure 6A1-Permeability vs T for 4F1

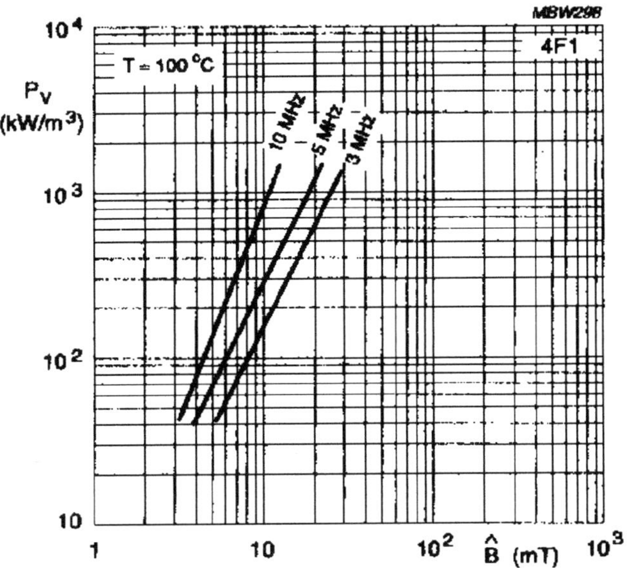

Figure 6A2- Core loss vs B for 4F1

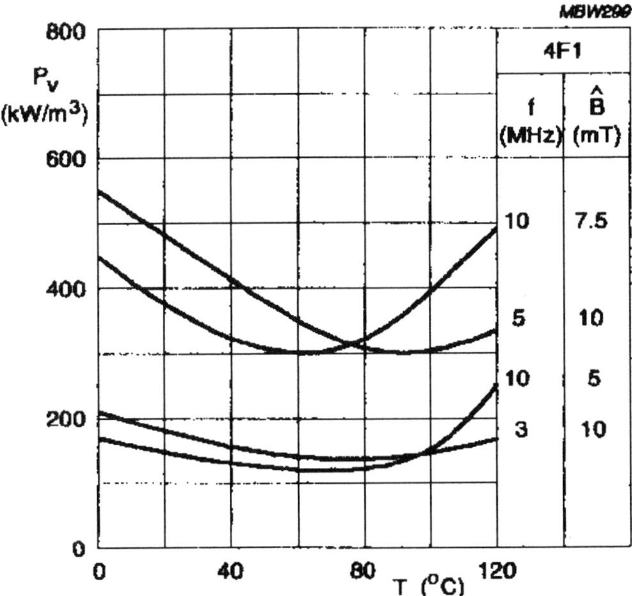

Figure 6A3- Core Loss vs T for 4F1

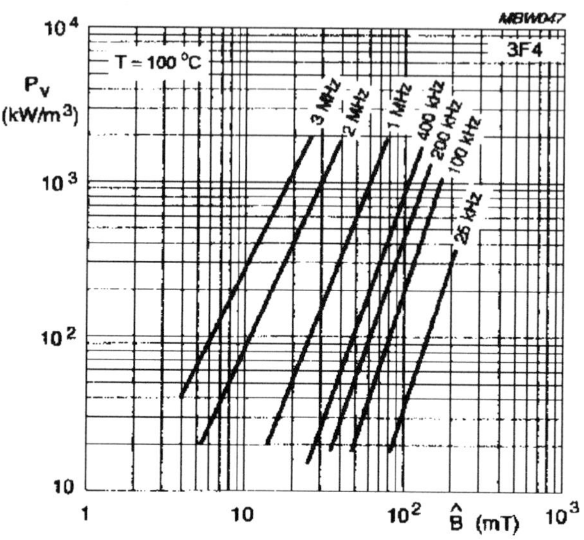

Figure 6A4- Core loss vs B for 3F4

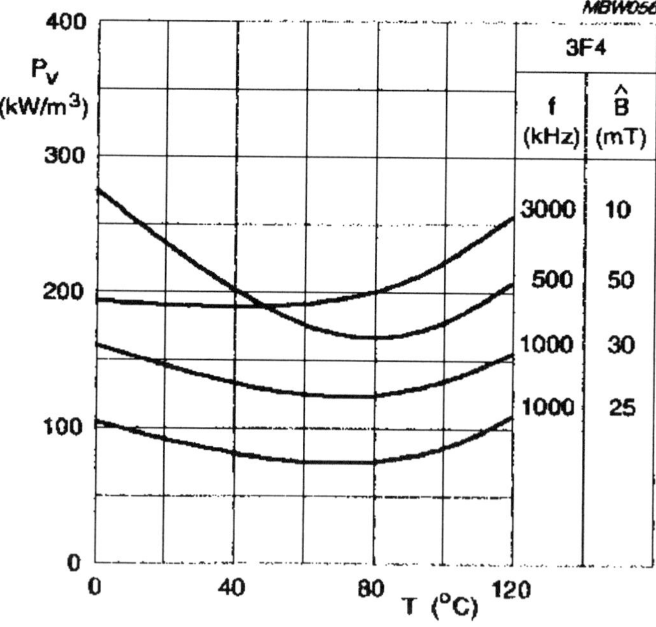

Figure 6A5-Core Loss vs T for 3F4

Table 6A17- Epcos Power Materials vs Application

Usage	Frequency range	Material	Specific application	Type
Power transformers, chokes	1 to 100 kHz	N 27	Transformers for flyback converters	E, EC, ETD, U, RM, PM
		N 41	Chokes	Pot cores, RM
	up to 200 kHz	N 53	Diode splitting transformers	E, U, UR
		N 62		E, U, UR, ETD, ER
		N 67	High-voltage transformers	
		N 72	Electronic lamp ballast devices	E, ETD
	up to 300 kHz	N 82	Diode splitting transformers	U, UR
	up to 500 kHz	N 87	Transformers for forward and push-pull converters	ETD, EFD, RM, TT/PR, ER, ELP
	0,3 – 1 MHz	N 49	Transformers for DC/DC converters, particularly resonance converters	EFD, ER, ELP, RM (low profile)
	0,5 – 1 MHz	N 59		

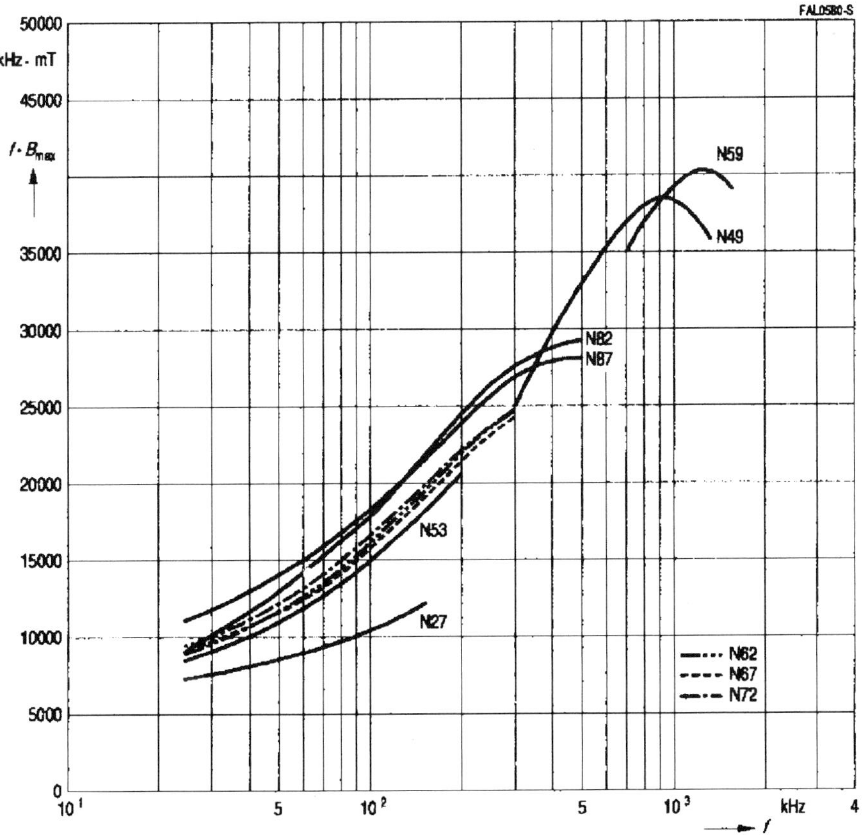

Figure 6A6- PF vs Frequency for Epcos Power Materials

Table 6A18-Epcos Power Materials

Preferred application			Power transformers				
Material			N 59	N 49	N 53	N 82[4)]	N 62
Base material			MnZn	MnZn	MnZn	MnZn	MnZn
	Symbol	Unit					
Initial permeability (T = 25 °C)	μ_i		850 ± 25 %	1300 ± 25 %	1700 ± 25 %	1900 ± 25 %	1900 ± 25 %
Flux density (H = 1200 A/m, f = 10 kHz)	B_S (25 °C) B_S (100 °C)	mT mT	460 370	460 370	490 420	490 415	500 410
Coercive field strength (f = 10 kHz)	H_c (25 °C) H_c (100 °C)	A/m A/m	60 50	55 45	26 16	17 11	18 11
Typical frequency range		kHz	500 … 1500	300 … 1000	16 … 200	16 … 300	16 … 200
Hysteresis material constant	η_B	10^{-6}/mT	—	—	—	—	—
Curie temperature	T_C	°C	> 240	> 240	> 240	> 240	> 240
Mean value of α_F at 20 … 55 °C		10^{-6}/K	—	—	—	—	—
Density (typical values)		kg/m³	4750	4750	4800	4800	4800
Relative core losses (typical values)	P_V						
25 kHz, 200 mT, 100 °C		mW/g mW/cm³			20 100	14 69	16 80
100 kHz, 200 mT, 100 °C		mW/g mW/cm³			125 625	84 421	105 525
300 kHz, 100 mT, 100 °C		mW/g mW/cm³		120 600	135 670	88 440	
500 kHz, 50 mT, 100 °C		mW/g mW/cm³	39 180	24 120			
1 MHz, 50 mT, 100 °C		mW/g mW/cm³	110 510	115 560			
Resistivity	ρ	Ωm	26	11	6	11	4
Core shapes			EFD	RM, Ring, EFD, ER ELP	E, U	U, UR	ETD, E, U
Other material properties (graphs) see page			78	81	84	87	90

Table 6A19-Epcos Power Materials

Material properties (continued)

Preferred application			Power transformers				
Material			N 27	N 67[5)]	N 87	N 72	N 41
Base material			MnZn	MnZn	MnZn	MnZn	MnZn
	Symbol	Unit					
Initial permeability (T = 25 °C)	μ_i		2000 ± 25 %	2100 ± 25 %	2200 ± 25 %	2500 ± 25 %	2800 ± 25 %
Flux density (H = 1200 A/m, f = 10 kHz)	B_S (25 °C) B_S (100 °C)	mT mT	500 410	480 380	480 380	480 370	490 390
Coercive field strength (f = 10 kHz)	H_c (25 °C) H_c (100 °C)	A/m	23 19	20 14	16 9	15 11	22 20
Typical frequency range		kHz	25 … 150	25 … 300	25 … 500	25 … 300	25 … 150
Hysteresis material constant	η_B	10^{-6}/mT	< 1,5	< 1,4	< 1,4	—	< 1,4
Curie temperature	T_C	°C	> 220	> 220	> 210	> 210	> 220
Mean value of α_F at 20 … 55 °C		10^{-6}/K	3	4	4	—	4
Density (typical values)		kg/m³	4750	4800	4800	4800	4800
Relative core losses (typical values)	P_V						
25 kHz, 200 mT, 100 °C		mW/g mW/cm³	32 155	17 80		16 80	35 180
100 kHz, 200 mT, 100 °C		mW/g mW/cm³	190 920	105 525	80 385	110 540	280 1400
300 kHz, 100 mT, 100 °C		mW/g mW/cm³		115 560	85 410		
500 kHz, 50 mT, 100 °C		mW/g mW/cm³					
1 MHz, 50 mT, 100 °C		mW/g mW/cm³					
Resistivity	ρ	Ωm	3	6	8	12	2
Core shapes			P, PM, ETD, EC, ER, E, U, Ring	RM, P, EP, ETD, ER, EFD, E, U, Ring	RM, TT, P, PM, ETD, EFD, E, ER, ELP	E, EFD	RM, P
Other material properties (graphs) see page			93	96	99	102	105

Table 6A20-Steward Common Mode and EMI Toroids

Material			21**	25**	26	28	24	27	28*	33
Relative Initial Permeability	μ_i		16	125	850	850	1050	1500	1700	2700
Tolerance		%			± 20				± 30	± 20
Saturation Flux Density	B_s	Gauss	2500	3600	2900	3250	2750	2800	3000	4700
		mT	250	360	290	325	275	280	300	470
at Field Intensity	H	Oersteds	80	10	10	10	10	10	10	10
		A/m	6400	800	800	800	800	800	800	800
Residual Flux Density	B_r	Gauss	1500	2600	1800	2000	1400	1200	1450	1200
		mT	150	260	180	200	140	120	145	120
Coercive Force	H_c	Oersteds	20	1.6	0.35	0.4	.30	.25		.20
		A/m	1600	127	28	3	24	20		16
Relative Loss Factor at Frequency	$\tan\delta/\mu_i$ f	10^{-6} MHz			30 .10		30 .10	50 .10		6 .10
Disaccommodation Factor		10^{-6}			< 10					< 2.5
Curie Temperature	T_c	°C	> 500	> 225	> 145	> 175	> 140	> 120	> 120	> 200
Resistivity	ρ	Ω-cm	10^7	10^6	10^7	10^5	10^6	10^6	10^5	5×10^2
Density		g/cm³	4.5	4.9	4.8	4.9	4.9	4.9	4.8	4.7

Broadband Common Mode Choke Cores

Specialty Material

Table 6A21-Steward Common Mode and EMI Toroids

Material			34*	31	36**	35	37	42***	40
Relative Initial Permeability	μ_i		4000	4500	5000	5000	7500	7500	10000
Tolerance		%	± 20	±20	±25	±20	±25	± 25	±30
Saturation Flux Density	B_s	Gauss	4200	4300	4200	4100	4100	4100	4000
		mT	420	430	420	410	410	410	400
at Field Intensity	H	Oersteds	10	10	10	10	10	10	10
		A/m	800	800	800	800	800	800	800
Residual Flux Density	B_r	Gauss	1200	1200		1300	1050	1100	1000
		mT	120	120		130	105	110	100
Coercive Force	H_c	Oersteds	.12	.10		.10	0.05	.08	0.04
		A/m	10	8		8	4	6	3
Relative Loss Factor at Frequency	$\tan\delta/\mu_i$ f	10^{-6} MHz	20 .10	20 .10	10 .10	20 .10	5 .010	6 .010	5 .010
Disaccommodation Factor		10^{-6}	< 2.5	< 2.5	< 2.5	< 2.0	< 2.5	< 3.0	< 2.5
Curie Temperature	T_c	°C	> 150	> 165	> 170	> 150	> 130	> 130	> 120
Resistivity	ρ	Ω-cm	10^2	10^2	10^2	10^2	10^2	10^2	50
Density		g/cm³	4.75	4.8	4.75	4.8	4.8	4.8	4.8

Table 6A22- Steward EMI Suppression Materials

Property Units	Frequency Range MHz	Initial Permeability (μ_i)	Saturation Flux Density (B_s) Gauss	Residual Flux Density (B_R) Gauss	Coercive Force H_c	Curie Temperature (T_c) °C	Volume Resistivity (ρ) Ohm-Centimeters
25 Material	100 - 500	125	3650 @ 10 Oe	2600	1.6	≥ 225° C	10^8
28 Material	30 - 300	850	3350 @ 10 Oe	2200	0.4	≥ 175° C	10^5
29 Material	30 - 300	600	3300 @ 10 Oe	1500	0.35	≥ 165° C	10^8

Steward's 25 and 29 Materials have higher resistivity than most other ferrites. They virtually eliminate cross talk on multiline filters and connector components.

- # 25 Material is designed for 200 MHz and up.
- # 28 Material is designed for 100 MHz.
- # 29 Material is designed to eliminate cross talk in multi-hole devices.

Table 6A23-Ferronics Common-Mode Materials

Characteristic	V	T	B	G	J	K*	P*	Units
Initial Permeability (μ_i)	15,000	10,000	5000	1500	850	125	40	
Loss Factor (tan δ/μ_i)	≤ 7	≤ 7	≤ 15	60		150	85	X 10^{-6}
at frequency =	0.01	0.01	0.1	0.1	0.1	10	10	MHz
Hysteresis Factor (h/μ^2)	-	-	< 2	10	6	-	-	X 10^{-6}
Saturation Flux Density (B_s)	3700	3800	4500	3200	2800	3200	2150	Gauss
at H max=	12.6	12.6	12.6	12.6	12.6	25	25	Oersted
Remanence (B_r)	1500	1400	1000	1500	1800	1600	400	Gauss
Coercivity (H_c)	0.03	0.04	0.07	0.25	0.4	1.5	3.5	Oersted
Curie Temperature (T_c)	≥ 120	≥ 120	≥ 160	≥ 125	≥ 120	≥ 350	≥ 350	°C
Temperature Coefficient of μ_i (α) -40° to +80°C (T.C.)	0.6	0.6	0.9	0.7	0.7	0.1	0.1	%/°C
Volume Resistivity (ρ)	25	40	≥ 10^2	≥ 10^6	≥ 10^5	≥ 10^7	≥ 10^8	Ω-cm

Table 6A24- FDK Suppession Materials

Material			K32	L51	K14	K26
Initial permeability	μiac	—	700	350	100	40
Saturation magnetic flux density	Bs	mT	300	340	360	350
Residual magnetic flux density	Br	mT	180	250	230	100
Coercivity	Hc	A / m	40	500	320	750
Relative loss factor	tan δ/μ	$\times 10^{-5}$	1.7	1.5	5.5	15
		MHz	0.4	0.5	6.0	10
Temperature coefficient	$\alpha\ \mu r$	$\times 10^{5}$	7	20	10	20
Curie temperature	Tc	°C	>160	>200	>270	>300
Resistivity	ρ	Ω · m	10^6	10^3	10^3	10^6
Density	d	$\times 10^3 Kg/m^3$	5.0	5.0	4.9	4.9

Property	Symbol	Condition	Unit	6H10	6H20	6H40	6H41	6H42	7H10	7H20
AC Initial permeability	μi	0.1 MHz	—	2500	2300	2400	2500	3400	1500	1000
Saturation magnetic flux density	Bs (1000 A/m)	23 °C	mT	510	510	530	530	530	480	480
		100 °C		390	390	430	430	430	380	380
Residual magnetic flux density	Br	23 °C	mT	110	130	110	110	110	150	130
Coercivity	Hc	23 °C	A/m	13	13	10	10	10	30	25
Relative loss factor	tanδ/μi	0.1 MHz	$\times 10^{-6}$	<5	<5	<3	<3	<3	<5	<4
Core loss	200 mT, 25 kHz	23 °C	kW/m³	—	—	90	75	60	—	—
		40 °C		—	—	75	60	50	—	—
		60 °C		65	80	60	50	40	—	—
		80 °C		55	65	50	40	45	—	—
		100 °C		80	55	40	45	55	—	—
	200 mT, 100 kHz	23 °C	kW/m³	—	—	650	550	450	—	—
		40 °C		—	—	550	450	350	—	—
		60 °C		450	550	450	350	300	—	—
		80 °C		400	450	350	300	325	—	—
		100 °C		500	400	300	325	375	—	—
	50 mT, 500 kHz	60 °C	kW/m³	—	—	—	—	—	100	50
		80 °C		—	—	—	—	—	80	40
		100 °C		—	—	—	—	—	100	50
	50 mT, 1 MHz	60 °C	kW/m³	—	—	—	—	—	400	200
		80 °C		—	—	—	—	—	400	200
		100 °C		—	—	—	—	—	500	250
Temperature coefficient	αμr	20 °C~80 °C	$\times 10^{-6}$	8	8	8	8	8	8	8
Curie temperature	Tc	—	°C	>200	>200	>200	>200	>200	>200	>200
Resistivity	ρ	—	Ω · m	3	3	2	2	2	5	5
Apparent density	d	—	$\times 10^3$ kg/m³	4.8	4.8	4.9	4.9	4.9	4.8	4.8

Note: 1) The values were obtained with toroidal cores (FR25/15/5).
2) The values were obtained at 23±2 °C unless otherwise specified.
3) Initial permeability was measured at 10kHz, 0.8A/m.

Table 6A25- FDK Power Ferrite Materials

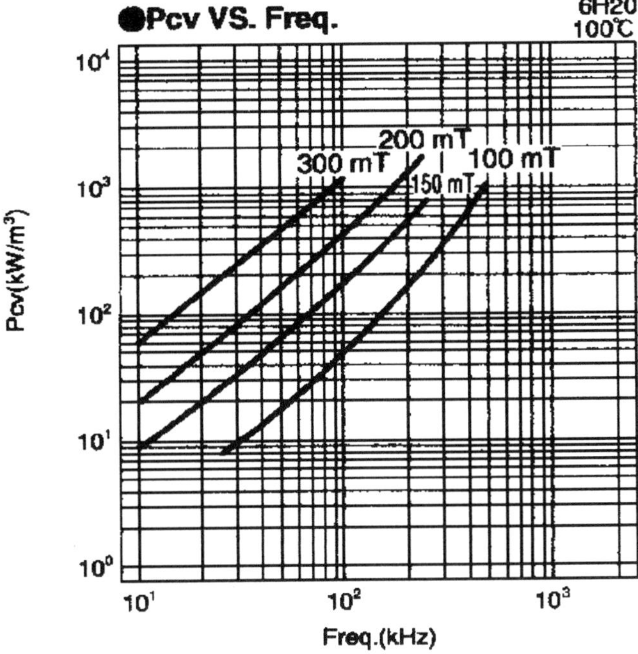

Figure 6A7- Core loss for 6H20

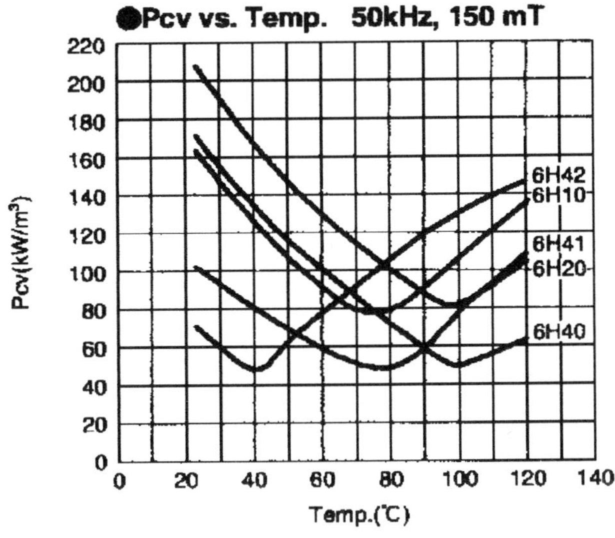

Figure 6A8- Core Loss vs T for FDK power ferrites

POWER APPLICATIONS

Symbols	Units	Test conditions	B1 PW1a / PW1b	B2 PW3b	B3 PW1b	B5 Standard PW2a / PW2b	B7 IEC 1332 PW2b	F1 PW3b	F2 PW4b	*F4 PW5b
μ_i		25°C	2500 ± 25%	1900 ± 25%	1900 ± 25%	1800 ± 25%	2000 ± 25%	2300 ± 25%	1900 ± 25%	1100 ± 25%
B at H (nominal values)	mT	400 A/m 25°C	450	460	470	470	470	450	420	390
		400 A/m 100°C	340	360	380	380	380	340	320	310
		1600 A/m 25°C	480	490	500	500	500	480	450	420
		1600 A/m 100°C	370	380	400	400	400	370	350	330
H_c	A/m	25°C	12	16	16	16	16	16	15	15
		100°C	10	10	10	10	10	10	10	10
T_c	°C		> 200	> 250	> 250	> 250	> 250	> 230	> 200	> 200
$P_{L\,typ}$	mW/cm³	16 kHz - 100°C 200 mT	< 100		< 80					
		25 kHz - 100°C 200 mT	< 180		< 150					
		32 kHz - 100°C 200 mT	< 250		< 200	< 140	< 120			
		60 kHz - 100°C 200 mT		< 340		< 350	< 330	< 280		
		100 kHz - 100°C 100 mT		< 150						
		100 kHz - 100°C 200 mT				< 700	< 680	< 580		
		300 kHz - 100°C 50 mT		< 120					< 100	
		500 kHz - 100°C 50 mT							< 230	< 180
		1 MHz - 100°C 50 mT								< 600
		1.5 MHz - 100°C 50 mT								< 1200
ρ	Ω x m		1	6	6	6	6	6	6	6
Density	g/cm³		4.8	4.8	4.8	4.8	4.8	4.8	4.6	4.6

Values measured on Ø 35 X Ø 12x18 reference toroid.
*Values measured on Ø 27.1 X Ø 13.8x11 reference toroid.

Table 6A26- Power Materials -AVX-Thomson

Table 6A27 AVX Common-Mode Filter Materials

FILTERING APPLICATIONS

Symbols	Units	Test conditions	A2 CL11	A3 CL10	A4 CL9	A5 Class CL9	A6 CL8	A8 CL8	A9 CL
μi		25°C	10000 ± 30%	7500 ± 25%	6000 ± 25%	5000 ± 25%	4000 ± 25%	3500 ± 25 %	2500 ± 25%
$\hat{B}$ at $\hat{H}$ (Typical values)	mT	25°C	330	330	350	350	410	480	480
		100°C	200	200	250	250	310	370	370
	A/m		800	800	800	800	800	1600	1600
H_c	A/m	25°C	6.2	6.2	6.4	6.4	12	12	12
		100°C	3.1	3.1	4.8	4.8	8	10	10
T_c	°C		> 120	> 120	> 140	> 140	> 160	> 200	> 200
f_c	MHz	25°C	0.3	0.3	0.3	0.5	0.6	1	1.5
tgδ /μ	x10-6	25°C	< 7	< 7	< 9	< 6	< 9		< 8
at f	kHz		10	10	10	10	10		100
ρ	Ω x m		0.3	0.3	0.5	0.5	0.5	2	1
Density	g/cm³		4.9	4.9	4.8	4.8	4.7	4.8	4.8

/alues measured on Ø 21 X Ø 14x10 reference toroid.

Table 6A28- Power Materials Kaschke

Werkstoff/*Material*	K 2004	K 2500	K 2006	K 4000	K 6000
Anfangspermeabilität μ_i *Initial permeability* μ_i	2000 ±25%	2500 ±25%	2100 ±25%	4000 ±25%	6000 ±25%
Induktion/*Induction* B_{max}/mT bei/*at*	455	≥ 490	490	380	370
Feldstärke/*Field strength* H_{max}/Am^{-1}	800	800	800	800	800
Remanenz/*Remanence* B_r/mT	200	≥ 200	200	120	150
Koerzitivfeldstärke/*Coercivity* H_C/Am^{-1}	20	≤ 20	≤ 20	≤ 10	≤ 8
Curietemperatur/*Curie temperature* ϑ_c/°C	≥ 200	≥ 220	≥ 200	≥ 130	≥ 130
Bezogener Temperaturbeiwert *Relative temperature factor* $\alpha_F \cdot 10^{-6}$/K^{-1} bei/*at* - 25...+25°C + 5...+25°C +25...+55°C + 25...+70°C	für Leistungsübertrager *for power transformers*	- - -2...+3 -	für Leistungsübertrager *for power transformers*	≤ 1,5 ≤ 1,0 ≤ 0,5 ≤ 0,5	≤ 2,0 ≤ 1,0 ≤ 1,0 ≤ 1,0
Schwerpunkt von /*Average value* $\alpha_F \cdot 10^{-6}$/K^{-1} bei/*at* 20 - 55°C	-	-	-	≤ 0,5	≤ 1,0
Bezogener Verlustfaktor *Relative loss factor* tan $\delta/\mu_i \cdot 10^{-6}$ bei/*at* 10 kHz 50 kHz 70 kHz 100 kHz	- - - -	≤ 2 - - ≤ 3	- - - -	≤ 4,2 ≤ 10 ≤ 17 ≤ 50	≤ 4 ≤ 12 ≤ 20 ≤ 30
Hysteresebeiwert/ *Hysteresis loss coefficient* $\eta_B \cdot 10^{-6}$/mT^{-1}	-	-	-	1,5	1
Verlustleistung/*Power loss* Pv W/cm³ bei/*at* 16 kHz 25°C 100°C	≤ 0,12 ≤ 0,10	≤ 0,07 ≤ 0,09	≤ 0,09 ≤ 0,065	- -	- -
Induktion/*Induction* B bei/*at* 100°C, 250 A/m, 16 kHz	≥ 330	-	≥ 330	-	-
Desakkommodationsfaktor *Disaccommodation factor* $D_F \cdot 10^{-6}$	-			≤ 6	≤ 4,5
Elektrischer Widerstand *Resistivity* ρ/Ω *m*	≥ 1	≥ 1	≥ 1	≥ 1	≥ 0,5

Table 6A29- Properties of Kaschke 2004

Anfangspermeabilität/Initial permeability μ_i		2000 ± 25%
Flußdichte/Flux density B bei Feldstärke/at field strength H		≥ 455 mT 800 A/m
Remanenz/Remanence B_r		≥ 200 mT
Koerzitivfeldstärke/Coercivity H_c		≤ 20 A/m
Curietemperatur/Curie temperature ϑ_c		≥ 200°C
Bez. Temperaturbeiwert/Rel. temperature factor $\alpha_F \cdot 10^{-6}$/K		
bei/at	-25 ... +25°C	≤ 6
	+5 ... +25°C	≤ 6
	+25 ... +55°C	≤ 10
	+25 ... +70°C	≤ 10
Bez. Verlustfaktor/Rel. loss factor $\tan\delta/\mu_i \cdot 10^{-6}$		
bei/at	10 kHz	≤ 12
	50 kHz	≤ 30
	70 kHz	≤ 40
	100 kHz	≤ 50
	200 kHz	≤ 100
	300 kHz	≤ 150
	500 kHz	≤ 300
Leistungsverluste/Power loss P_v		
bei / at	16 kHz, 200 mT, 25°C	≤ 120 mW/cm^3
	16 kHz, 200 mT, 100°C	≤ 100 mW/cm^3
Induktion/Induction B bei/at 100°C, 16 kHz, 250 A/m		≥ 330 mT
Spez. Widerstand/Resistivity ρ		≥ 1 Ωm

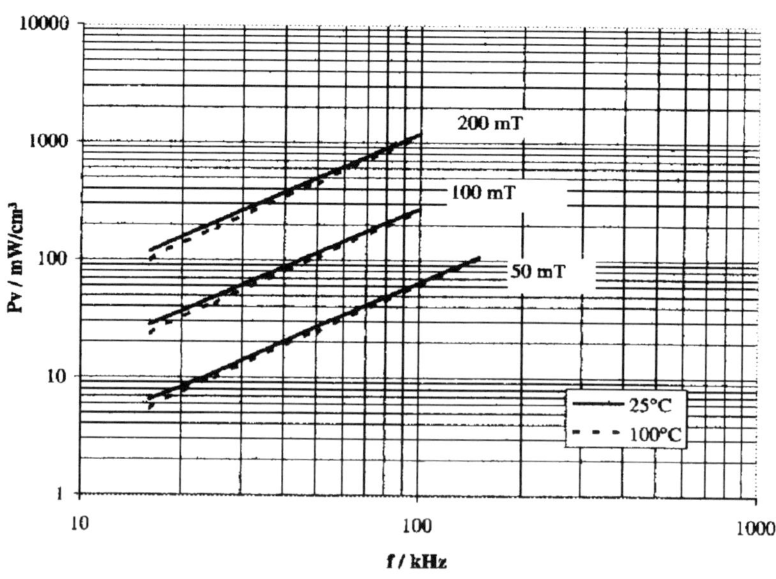

Figure 6A9- Core Loss of Kaschke 2004

Materials 材質名	10KHz	10KHz	20°C~60°C	1000A/m (mT)	(A/m)	(°C)	(Ω-m)	(kg/m³)	
2H1	15000	0.95×10^{-5}	0.17×10^{-6}/°C	350	2.3	105	0.06	5.0×10^{3}	
2E2	10000	1.5×10^{-5}	0.1×10^{-6}/°C	370	4.0	120	0.024	4.9×10^{3}	3
2E2B	10000	0.34×10^{-5}	0.28×10^{-6}/°C	390	6.0	120	0.047	4.9×10^{3}	
2E1	7000	1.8×10^{-5}	-0.8×10^{-6}/°C	415	8.0	150	0.012	4.9×10^{3}	3
2G1	7000	0.54×10^{-5}	0.41×10^{-6}/°C	410	6.0	135	0.02	4.8×10^{3}	
2G3	6000	0.16×10^{-5}	0.28×10^{-6}/°C	430	6.7	145	0.2	4.9×10^{3}	
2F1	5300	0.2×10^{-5}	1.0×10^{-6}/°C	370	7.2	120	0.13	4.8×10^{3}	4
2D3C	4000	0.1×10^{-5}	0.2×10^{-6}/°C	410	8.0	140	0.1	4.9×10^{3}	
2E4	4000	0.15×10^{-5}	-0.5×10^{-6}/°C	450	11.9	180	0.1	4.8×10^{3}	
2F6	3300	0.07×10^{-5}	2.65×10^{-6}/°C	470	11.9	>200	2.1	4.8×10^{3}	5
2D3	3000	0.3×10^{-5}	-0.4×10^{-6}/°C	450	11.9	160	0.9	4.9×10^{3}	
2E6	3000	0.1×10^{-5}	-0.5×10^{-6}/°C	490	11.9	>200	0.23	4.9×10^{3}	4
2E7	2400	0.15×10^{-5}	1.25×10^{-6}/°C	490	11.9	190	3	4.8×10^{3}	
2F8	2200	0.1×10^{-5}	6.3×10^{-6}/°C	490	9.5	>200	8	4.8×10^{3}	5
2E6C	2000	0.16×10^{-5}	7.5×10^{-6}/°C	510	14.3	>230	0.37	4.8×10^{3}	
2C3	2000	0.6×10^{-5}	3.0×10^{-6}/°C	370	15.9	110	1.3	4.9×10^{3}	
2H6	800	0.13×10^{-5} (500KHz)	0.4×10^{-6}/°C	490	22.0	>200	2.3	4.8×10^{3}	

Table 6A30- Tomita Power Materials

Table 6A31-Vogt Power Materials

FERROCARIT			Fi 325 [1]	Fi 324	FI 323	FI 322 [2]	FI 311 [3]
Anfangspermeabilität *Initial permeability*	μi	1	1800 ± 25%	2300 ± 25%	2500 ± 25%	1600 ± 25%	1400 ± 25%
Bezogener Verlustfaktor *Relative loss factor*	$\frac{\tan\delta}{\mu_i}$	10^{-6}			< 6	< 6	< 2,5 / < 10
Frequenz *frequency*	f	MHz			0,1	0,05	0,01 / 0,2
Hysterese - Konstante *Hysteresis material constant*	η_B	$\frac{10^{-6}}{mT}$			< 1		< 1,5
Induktion *Induction*	$\hat{B}$ bei $\hat{H}$ = 3000 A/m	mT	510	490	490	490	390
Koerzitivfeldstärke *Coercivity*		A/m	16	20	20	20	30
Curie - Temperatur *Curie temperature*		°C	230	230	230	230	220
Bez. Temperatur - Beiwert *Rel. temperature factor* + 23 ... + 70°C - 20 ... + 23°C	α_F	$\frac{10^{-6}}{K}$			1 ... 2		< 5
Bezogener Desakkommo-dationsbeiwert *Rel. disaccommodation factor*	D_F bei *at* 40°C	10^{-6}			< 3		
Spezif. Gleichstrom-Widerstand *DC - Resistivity*		Ωm	≥ 6	≥ 3	> 0,5	> 1	> 3

Table 6A32- Domen Ferrite Materials

Specs	Unit	Mn-Zn ferrites		Ni-Zn ferrites	
		2500 HMC1	2500 HMC2	250BHC	300BHC
μi	—	4500 (+ 20 °C) 4100 (+120 °C)	4500 (+ 20 °C) 4100 (+120 °C)	250	300
fc	mHz	0.4	—	8	8
Psp	$\frac{mW}{cm^3Hz}$	10.5 (+25 °C) 8.7 (+100 °C)	8.5 (+25 °C) 6.0 (+100 °C)	30 (3)	<30 (3)
$tg\delta_{\mu/\mu i}$	10^{-6} (MHz)	—	—	30 (3)	<30 (3)
B	mT	290	330	—	320
Bm	mT	450	—	—	130
Br	mT	100	—	—	80
Hc	A/m	16	—	—	—
Tc	°C	>200	>200	>250	>250

Table 6A33- Vogt Common-Mode Materials

FERROCARIT			Fi 410		Fi 360		Fi 350		Fi 340	
Anfangspermeabilität *Initial permeability*	μ_i	1	10000 ± 30%		6000 ± 20%		5000 ± 20%		4300 ± 20%	
Bezogener Verlustfaktor *Relative loss factor*	$\frac{\tan\delta}{\mu_i}$	10^{-6}	< 6	< 70	< 4	< 20	< 4	< 20	< 4	< 20
Frequenz *frequency*	f	MHz	0,01	0,1	0,01	0,1	0,01	0,1	0,01	0,1
Hysterese - Konstante *Hysteresis material constant*	η_B	$\frac{10^{-6}}{mT}$	< 0,6		< 0,8		< 0,8		< 1	
Induktion *Induction*	$\hat{B}$ bei *at* $\hat{H}$ = 3000 A/m	mT	400		440		420		390	
Koerzitivfeldstärke *Coercivity*		A/m	8		6		10		10	
Curie - Temperatur *Curie temperature*		°C	130		150		150		130	
Bez. Temperatur - Beiwert *Rel. temperature factor* + 23 ... + 70°C − 20 ... + 23°C	α_F	$\frac{10^{-6}}{K}$	≤ 3,0		≤ 1,5		< 1,5		1 ... 2	
Bezogener Desakkommodationsbeiwert *Rel. disaccommodation factor*	D_F bei *at* 40°C	10^{-6}	< 3		< 3		< 3		< 6	
Spezif. Gleichstrom-Widerstand *DC - Resistivity*		Ωm	≥ 0,05		> 0,05		> 0,1		> 0,5	

Table 6A34-Iskra Power Materials

Material			2E	1E	1F	3F	3C	2C	10G	1C	5G	15G	25G	35G
μ_i			13 + 20 %	25 ± 20 %	80 ± 20 %	125 ± 20 %	200 ± 20 %	300 ± 20 %	750 ⊥ 20 %	900 ± 20 %	2 200 ± 20 %	2 200 ± 20 %	3 000 ± 20 %	1 400 ± 20 %
$\tan\delta/\mu_i$	pri/at	10^{-6}	< 350	< 150	< 50	< 35	< 45		< 12	< 10				
		MHz	30	3	3	3	1		0,3	0,01				
$\tan\delta/\mu_i$	pri/at	10^{-6}	< 1 000	< 600	< 65	< 70	< 45	< 20	< 22					
		MHz	100	40	10	10	3	1	1	0,1				
η_B	$\hat{B}$ = 1,5 in/and 3 mT	10^{-3}/T							< 2					
D_F	10 in 100 min. po razm. 10 and 100 min. after demagnet.	10^{-6}							< 8					
α_F	25 do/to 55°C	10^{-6}/K		3 do/ /to 14	2 do/ /to 6	0 do/ /to 8	0 do/ /to 8	0 do/ /to 10	0,5 do/ /to 2,5	0 do/ /to 4				
	5 do/to 25°C			− 0,5 do/ /to 14	1 do/ /to 6				0,5 do/ /to 2,8					
	− 25 do/to 25°C			− 1 do/ /to 14	1 do/ /to 6				0,5 do/ /to 3,2					
ϑ_c		°C	> 300	> 300	> 300	> 350	> 300	> 200	> 200	> 145	> 200	> 200	> 230	> 200
$\hat{B}$	$\hat{H}$ − 3 000 A/m	mT	80	120	320	370	300	300	400	350	450	500	470	420
H_c		A/m	1 500	1 100	400	250	150	100	100	50	20	20	20	30
$\hat{B}$	f − 16 kHz; ϑ − 100°C H − 250 A/m	mT									≥ 290	> 300		

Table 6A35-Hitachi Power Materials

CHARACTERISTICS \ MATERIALS	SYMBOL	UNIT	TEMP	SB-5S	SB-3L	SB-7C	SB-9C
AC INITIAL PERMEABILITY	μiac	--	23℃	3,000 ±25%	1,600 ±25%	2,400 ±25%	2,600 ±25%
SATURATION FLUX DENSITY	Bms	mT	23℃	480	510	500	490
			100℃	340	420	380	360
RESIDUAL FLUX DENSITY	Brms	mT	23℃	180	180	150	140
			100℃		100	70	60
COERCIVE FORCE	Hcms	A/m	23℃	12	15	13	12
			100℃		8	7	6
CORE LOSS (f=25kHz Bm=200mT)	Pc	kW/m³	23℃	140	200	140	145
			60℃	145	150	100	90
			100℃	175	120	80	70
			120℃		130	90	80
CORE LOSS (f=100kHz Bm=200mT)	Pc	kW/m³	23℃	1,100	1,100	780	680
			60℃	1,150	820	560	450
			100℃	1,500	750	500	400
			120℃		850	570	480
RELATIVE LOSS FACTOR	tanδ/μiac	$\times 10^{-6}$	23℃	※1 2.5	※1 3.5	※1 2.0	※2 5.0
CURIE TEMPERATURE	Tc	℃	—	>190	>260	>200	>200
SPECIFIC RESISTIVITY	ρ	Ω-m	23℃	0.4	1.5	5.0	5.0
DENSITY	ds	kg/m³	23℃	4.85×10^3	4.8×10^3	4.8×10^3	4.8×10^3

※1 f=10kHz

Table 6A36- Hitachi EMI Suppression Materials

CHARACTERISTICS \ MATERIALS	SYMBOL	UNIT	DL-2	DL-3	DL-4C	DL-5C	DL-6C	DL-7C	DL-8C
AC INITIAL PERMEABILITY	μiac	—	1,500	2,200	250	350	650	900	1,200
SATURATION FLUX DENSITY	Bms (B800)	mT	280	260	410	400	380	370	340
RESIDUAL FLUX DENSITY	Brms	mT	100	140	290	290	280	260	150
COERCIVE FORCE	Hcms	A/m	16	10	72	64	27	24	18
RELATIVE LOSS FACTOR	tanδ/μiac	$\times 10^{-6}$ at 100kHz	20	15	23	20	13	12	12
RELATIVE TEMP. FACTOR	$\alpha\mu^{r}$	$\times 10^{-6}$/°C	1.5	2	22	14	16	14	5

Table 6A37- TSC International Ferrites

Material Grade	Initial Permeability μ_0	Saturation Flux Density @15 oersteds B_s (Gauss)	Curie Temperature Tc (°C)	
TSF-5099	2,000	5,000	>210	Lowest core loss
TSF-7099	2,000	5,000	>210	For high ambient temperature applications
TSF-7070	2,200	5,000	>210	For potted applications
TSF-8040	3,100	5,100	>210	All purpose material for integrated magnetics
TSF-5000	5,000	4,300	>170	For filter inductors
TSF-010K	10,000	4,300	>125	For low harmonic distortion
TSF-Boost	2,000	5,000	>210	For dc bias applications

Table 6A38- Ceramic Magnetics Ferrites

		Pg.	8	9	10	12	14	16	18
Property	Symbol	Unit	MN30	MN60LL	MN60	MN67	MN80	MN100	MN8CX
INITIAL PERMEABILITY	μi	–	4000	6500	6500	1250	2400	10,000	2900
MAXIMUM PERMEABILITY*	μm	–	7000	10,500	10,500	8800	6000	15,500	3500
MAXIMUM FLUX DENSITY	B_m	gauss	4600	4500	4500	5200	5000	4500	4700
REMANENT FLUX DENSITY	Br	gauss	1300	800	800	2900	1500	700	700
COERCIVE FORCE	Hc	oersted	0.13	0.08	0.08	0.21	0.17	0.07	0.18
CURIE TEMPERATURE	Tc	°C	175	185	185	280	230	170	185
dc VOLUME RESISTIVITY	ρ	ohm-cm	400	500	200	150	200	150	1000
LOSS FACTOR X 10^6	1/μiQ	–							
@ 10 KHz			2	2.5	3	20	2.2	2	1.9
100 KHz			7	4	12	22	4.5	15	2.4
300 KHz			35	20	55	35	13	100	12
1 MHz			/	/	/	200	260	/	140
5 MHz			/	/	/	/	/	/	/
10 MHz			/	/	/	/	/	/	/

* @ 40 oersteds. Please consult factory for information on effects of field strength on magnetic properties of these materials.

Table 6A39-Samwha Power Ferrites

Materials				PL-5	PL-7	PL-9
Initial permeability	¥ìiac			2200¡¾25%	2400¡¾25%	3000¡¾25%
Core loss (100kHz, 200mT)	Pcv	nW/§¨	23¡É	800	650	450
			80¡É	550	450	350
			100¡É	500	410	390
Saturation flux density (1194A/m)	Bs	mT	23¡É	500	490	500
			100¡É	390	380	380
Remanence	Br	mT	23¡É	180	150	150
Coercivity	Hc	A/m	23¡É	15	12	10
Curie Temperature	Tc	¡É		>220	>210	>200
Density	¥ä	kg/§©		4.85x10^3	4.85x10^3	4.85x10^3
Resistivity	¥ñ	¥Ø-m		3	5	7
Initial permeability	¥ìiac			2200¡¾25%	2400¡¾25%	3000¡¾25%
Core loss (32kHz, 100mT)	Pcm	W/kg	23¡É	95	70	53
			80¡É	65	55	37
			100¡É	60	45	43
Saturation flux density (1194A/m)	Bs	mT	23¡É	500	490	500
			100¡É	400	380	380
Remanence	Br	mT	23¡É	180	150	130
Coercivity	Hc	A/m	23¡É	15	12	10
Curie Temperature	Tc	¡É		>220	>210	>200
Density	¥ä	kg/§©		4.85x10^3	4.85x10^3	4.85x10^3
Resistivity	¥ñ	¥Ø-m		3	5	7

Materials			SN-20	T-314	SN-065	SN-201
Initial permeability	¥ìiac		2000¡¾20%	1000¡¾20%	650¡¾20%	500¡¾20%
Relative loss factor	tan¥ä/¥ìiac	x10^6	25 (0.1Mhz)	30 (0.1Mhz)	30 (0.7Mhz)	30 (0.8Mhz)
Saturation flux density (1194A/m)	Bs	mT	260	280	300	230
Remanence	Br	mT	100	100	160	140
Coercivity	Hc	A/m	12	24	24	40
Relative loss factor (20¡É~60¡É)	¥á¥ì¥ä	x10^6/¡É	3~5	4~6	5~10	15
Curie Temperature	Tc	¡É	>100	>120	>150	>130
Density	¥ä	kg/§©	4.85x10^3	4.85x10^3	4.85x10^3	4.85x10^3
Resistivity	¥ñ	M¥Ø-m	>1.0	>1.0	>10	>10

Table 6A40-Cosmo Ferrites

PROPERTY		MATERIAL						
		CF 102	CF 138	CF 129	CF 196	CF 101	CF 195	CF197
INTIAL PERMEABILITY		500	2100	1900	2000	3000	5000	7500
SATURATION FLUX DENSITY	Bs (H=1KA/m)	220	480	510	500	490	400	400
		150	380	410	400	390	260	260
RESIDUAL FLUX DENSITY		150	180	180	210	200	150	150
COERCIVITY		45	15	15	16	15	12	12
POWER LOSS DENSITY	P_c (16k Hz) 200 mT	-	-	95	120	100	-	-
		-	-	60	110	120	-	-
	P_c (25k Hz) 200 mT	-	500	140	160	150	-	-
		-	80	95	140	170	-	-
RELATIVE LOSS FACTOR		-	2.5	2.5	4	2.5	5	7
SEC. MAX. PERMEABILITY		-	90-110	90-110	70-90	50-70	-	-
CURIE TEMPERATURE		>130	>220	>240	>200	>190	>120	>120
RESISTIVITY		1X10^6	1.0	1.0	0.4	0.4	0.2	0.2
DENSITY		4.4X10^6	4.8X10^3	4.8X10^3	4.8X10^3	4.8X10^3	4.85X10^3	4.85X10^3
GEOMETRY		B/W DY	EFD,EPC,EE, EI,ETD,EER,EC UU	EFD,EPC,EE, EI,ETD,EER,EC UU	EE,EI,ETD EER,EC,UU, TOROID	EE,EI,EER, UU	Small (EE,UU) TOROID	Small (EE,UU) TOROID

Table 6A41- Isu Ferrites

SYMBOLS	UNITS	TEST CONDITIONS	YM1	YM2	YM3	PM1	PM2	PM3	PM5
¥ìi		25.0¡É	900¡¾20%	350¡¾20%	480¡¾20%	2500¡¾20%	2000¡¾20%	1900¡¾20%	2200¡¾2
B		25.0¡É	330	250	280	450 | 480	470 | 500	480 | 510	480 | 5
at	mT	100¡É	250	190	190	340 | 370	380 | 400	390 | 410	390 | 4
H	A/m		1600	1600	1200	400 | 1600	400 | 1600	400 | 1600	400 16
Hc	A/m	25¡É	16	40	24	12	16	16	15
		100¡É	12	25	15	10	10	10	9
Tc	¡É		£¾150	£¾150	£¾150	£¾200	£¾250	£¾250	£¾25
fc	MHz	25¡É				1.8	2	2.5	2.5
PL	mW/	200mT85¡É	58* | 140*	95* | 200*	150* | 350*	130 | 130	75 | 115	105 | 240	210 | 5
		fKHZ	16 | 32	16 | 32	32 | 64	25 | 50	16 | 25	25 | 50	50 | 1C
¥ò	§Ù-m		50	£¾10	£¾10	1	3	5	6
Density	g/cm3		4.7	4.4	4.5	4.8	4.8	4.8	4.8
Core Shapes			Color Yoke-Cores	Color Yoke-Cores	Color MONITOR Yoke-Cores	E.U -Cores	E.U -Cores	E.U -Cores	E.U -Cores

SYMBOLS	UNITS	TEST CONDITIONS	PM1	PM2	PM2A	PM5	PM7	PM9
¥ìi		25¡É	2500¡¾20%	2000¡¾20%	2400¡¾20%	2200¡¾20%	2400¡¾20%	3000¡¾20%
B	mT	25¡É	450 | 480	470 | 500	480 | 480	480 | 510	460 | 480	480 | 500
at		100¡É	340 | 370	380 | 400	340 | 370	390 | 410	380 | 400	400 | 420
H	A/m		400 |1600	400 |1600	400 |1600	400 |1600	400 |1600	400 |1600
Hc	A/m	25¡É	12	16	12	15	12	11
		100¡É	10	10	9	9	8	7
Tc	¡É		£¾200	£¾250	£¾250	£¾210	£¾210	£¾200
fc	MHz	25¡É	1.8	2.0	1.8	1	1	1
P_L	mW/cm^3	200mT 85¡É	130 | 310	75 | 115	130 | 310		370	340
		200mT 100¡É				180 | 500	440	380
		f kHz	25 | 50	16 | 25	25 | 50	50 |100	100	100
ρ	§Ù-m		1	3	1	6	6	7
Density	g/cm^3		4.8	4.8	4.8	4.8	4.7	4.8
Core			E.U	E.U	E.U	E.U	E.U	E

Table 6A42-Acme-Maylaysia Ferrites

Type	Units	H2	H3	H4	H5	D1C	D28
f-Band MHz		0.5 - 35	0.5 - 20	0.1 - 3	0.2 - 7	0.4 - 5	0.1 - 3
μi		50 ± 25%	100 ± 25%	300 ± 25%	250 ± 25%	300 ± 25%	870 ± 25%
Bms	Gauss	3300	2700	2500	3300	3300	3100
	Oe	50	50	60	50	40	50
Br	Gauss	2100	1600	1500	850	1800	1700
Hc	Oe	5.52	1.19	0.43	0.52	0.45	0.65
tan δ/μi	$x10^{-6}$	180	50	20	110	30	20
	MHz	2	2	0.3	1	0.2	0.3
αμr	$x10^{-6}$/°C	50	37	1.5	30	3.8	35
T°c		350	250	150	220	160	165
ρ		10^7	10^7	10^7	10^6	10^6	10^7

Table 6A43- Acme Power Ferrites

	Symbol	Unit	Measuring Conditions			Low Loss Material			
			Freq.	Flux den.	Temp.	P2	P4	P5	S3
Initial Permeability	£g		10kHz	<0.1mT	25¡CC	2800±25%	2500±25%	2000±25%	≅ 800
Amplitude Permeability	£g		25kHz	200mT	25¡CC	>3200	>3200		
					100¡CC	>3200	>3200		
Power Loss	Pv	Kw/m3	25kHz	200mT	25¡CC	130	102		
					100¡CC	100	55		
			100kHz	200mT	25¡CC	900	700	750	
					100¡CC	700	450	600	1900
			300kHz	100mT	25¡CC		660	500	
					100¡CC		430	380	
			500kHz	50mT	25¡CC			190	
					100¡CC			150	
Saturation Flux Density	Bms	mT	1kHz	H=1000A/m	25¡CC	450	480	480	
					100¡CC	350	380	380	
Remanence	Brms	mT	1kHz	H=1000A/m	25¡CC	60	100	130	
					100¡CC	40	50	60	
Coercivity	Hc	Oe	1kHz	H=1000A/m	25¡CC	7	8	14	
					100¡CC	1.5	5	9	
Curie Temperature	Tc	¡CC				>200	>220	>220	
Resistivity	ρ	£[m				0.23	0.25	0.30	
Density	d	kg/m3				4.80	4.80	4.70	
Squareness			25kHz	H=1000A/m	25¡CC				

Table 6A44- Hinoday Ferrites

MATERIAL CHARACTERISTICS								
Parameters	**Symbol**	**Unit**	**Condition @**	**MSB-5S**	**MSB-7C**	**MSB-5F (H)**	**MGQ5C**	**MGP-9**
Ac Initial Permeability	µiac	---	25°C	3000±25%	2400±25%	1800±25%	5300±25%	7000±25%
Saturation Flux Density	Bm	mT	25°C	480	500	500	400	430
		Gauss	25°C	4800	5000	5000	4400	4300
Residual Flux Density	Br	mT	25°C	180	150	160	100	120
		Gauss	25°C	1800	1500	1600	1000	1200
Coercive Force	Hc	A/m	25°C	11.9	12.7	15.9	8.0	8.0
		Oe	25°C	0.15	0.16	0.20	0.10	0.10
Core Loss	Pc	mw/cc	16KHz/150mT/100 25KHz/200mT/100 100KHz/200mT/100 16KHz/200mT/40	— — — 75 Max.	40 Max. 95 Max. 600 Max. —	60 Max. --- --- ---	--- --- --- ---	--- --- --- ---
Relative Loss Factor	tanδ/µiac x 10^{-6}		0.01 MHz 0.1 MHz	<=2.5 —	<=2.0 ---	<=4.0 ---	— <=15	<=5.0
Curie Temperature	Tc	°C	---	>210	>220	>220	>150	>130
Density	Ds	gm/cc	---	4.85	4.85	4.85	4.85	4.85

Table 6A45-Mianyang Ferrites

	Unit	Materials		
		R2KD	R2KH	R2KBP1
Initial permebility ui		2000¡À25%	2000¡À25%	2000¡À25%
Effective saturation magnetic flux density Bs	mT	at 25C 480	at 25C 510	at 25C 510
		at 100C 380	at 100C 300	at 100C 390
Effective retentivity Br	mT	120	180	116
Effective coercivity Hc	A/m	16	18	15
Power losses P0	mw/g	at 16KHz,150mT 25C 12	at 16KHz,150mT 100C 5	at 25KHz,200mT 25C 27
		at 16KHz,150mT 100C 10	at 32KHz,200mT 100C 23	at 25KHz,200mT 60C 19
			at 64KHz,200mT 100C 60	at 25KHz,200mT 100C 20
Curie temperature Tc	C	>200	>200	>230
Electrical resistivity P	Ò.cm	100	300	1000
Apparent density d	g/cm3	4.8	4.8	4.8

Table 6A46-Hebei Ferrites

Characteristics	Symbol	Unit		
			R2KW	R2KPP
Frequency	f	MHZ	<0.4	<0.4
Initial pereability	¦Ìi	-	2000 ¡À20%	2500 ¡À25%
Relative loss factor	tan¦Ä/¦Ìi	10^{-6}	<2.5	-
Relative temperature coefficent	¦ÁF	10^{-6} / ¡ãC	0¡«1.5 (0¡«60¡ãC)	-
Flux density	Bs	mT	380	510
Remanence	Br	mT	80	117
Coercivity	Hc	A/m	20	12
Disac.factor	DF	10^{-6}	¡Ü5	-
Power loss 25kHz200mT	P_L	mW/cm3 100¡ãC	-	100
Curie temp.	Tc	¡ãC	>160	>200
Resistivity	¦Ñ	¦¸·m	1	10
Density	d	g/cm3	4.7	4.8

6.27-Magnetics Metallic Magnetic Strip Material

Magnetics lists tape wound cores for from conventional, and amorphous metallic materials. Table 6A47 gives the materials and Table 6A48 lists the gages into which cores are made. In figures are plotted the core loss versus flux density and frequency. The core loss units, unlike ferrites are given in Watt per pound.

6.28-Arnold Metallic Magnet Strip Material

Arnold Engineering also manufactures tape cores out of high frequency strip materials. The core losses for their materials are shown in Figure 6A10 for Silectron (SiFe). Figure 6A11 for 4-79 Permalloy, 6A12 for Supermalloy, 6A13 for Detamax (50 Permalloy) and 6A6A14 for Supermendur 50-50 CoFe.

6.29- Honeywell Amorphous Metallic Magnetic Strip Material

Honeywell took over manufacture of the amorphous metal alloy Metglas^R from Allied-Signal. They also produce the magnetic cores from the various alloys. The details of the Metglas® materials are to be found in Chapter 3 under Sections 3.3 with the properties listed under Table 3.2 and Figures 3.23-3.28. Cores from amorphous metal alloys are manufactured by Honeywell, Magnetics, Arnold, Vacuumschmelze and Coremaster.

6.30-Vacuumschmelze Amorphous and Nanocrystalline Alloys

Vacuumschmelze manufactures materials and cores out of amorphous and nanocrystalline alloys. This topic is discussed in Chapter 3 under Section 3.3 with properties under Table 3.2 and Figures 3.29-3.35 and also in Table 6A51 in the following appendix. Cores from VAC's materials are also sold by Coremaster International. (See Figures 6A18-24 and Table 6A49)

6.31-Toshiba Amorphous Alloys

Toshiba was probably the first company to commercialize cores from nanocrystalline materials. This topic is described in Chapter 3 under Section 3.42 with properties described in Table 3.4. Other properties are found in the following appendix at Table 6A49 and Figures 6A15-17.

6.32-TDK Amorphous Metal Cores

TDK manufactures cores from amorphous metal alloys. Table 6A52 lists the properties of their materials.

6.33- Coremaster Amorphous and Nanocrystalline Cores

Coremaster International manufactures cores from VAC's materials. See Figure 6A18. The materials are E-1000 S, E-2000 Q, Gapped E-2000 Q and E-5000 Y. Design of a push-pull output inductor using an E-2000 Q core is found in Appendix 5.6 in Chapter 5.

6.34-Magnetec Nanocrystalline Cores

Magnetec produces a nanocrystalline material calles Nanoperm®. The properties of cores from this material are listed in Figures 19-23 and Table 6A450.

6.35- Magnetics Powder Core Materials

Magnetics Powder cores are described in Chapter 3 under Section 3.52 with the properties for their Hi-Flux NiFe cores given in Figures 3.41-6. For their Kool-Mu (Sendust) cores, the topic is found in Chapter 3 under Section 3.53 with properties in Figures 3.47-3.50.

6.36-Arnold Engineering Powder Core Materials

Arnold makes the MPP and High-Flux cores as well as the Super MSS (Sendust) cores. Properties for the Hi-Flux cores are found in Figures 6A24- 6A27. The properties for their Super MSS are in Figures 6A28-6A32.

6.37-Micrometals Powdered Iron Cores

Properties for Micrometals powdered iron cores are found in Tables 6A53-55(Appendix 6.2) and in Figures 3.36-3.39(Chapter 3, Section 3.51

6.38-Pyroferric Powdered Iron Cores

Properties for Pyroferric cores are in Figures 6A33-37, Table 6A56 and Figure 3.40 in Chapter 3

Appendix 6.2- Listing of Catalog Data for Suppliers of Metallic Strip and Powder Cores

Tables Appendix List No.	Vendor	Figures Appendix List No.
6A47	Magnetics	
6A48	"	
	Arnold	6A10
	"	6A11
	"	6A12
	"	6A13
	"	6A14
6A49	Toshiba	6A15
	"	6A16
	"	6A17
	CoreMaster	6A18
	Magnetec	6A19
	"	6A20
6A50		6A21
		6A22
		6A23
	Arnold	6A34
	"	6A25
	"	6A26
	"	6A27
	"	6A28
	"	6A29
	"	6A30
	"	6A31
	"	6A32
6A51	Vacuumschmelze	
6A52	TDK	
6A53	Micrometals	

6A54	"	
6A55	"	
6A56	Pyroferric	6A33
	"	6A34
	"	6A35
	"	6A36
	"	6A37

Table 6A47- Metal Strip Materials for Power Electronics

MAT'L CODE	MATERIAL TYPE	FLUX DENSITY (KILOGAUSSES)	FLUX DENSITY (TESLAS)	SQUARENESS	COERCIVE FORCE DC OERSTEDS	COERCIVE FORCE DC A/M	COERCIVE FORCE 400 HERTZ CCFR** OERSTEDS	COERCIVE FORCE 400 HERTZ CCFR** A/M	GAIN***
K	Magnesil	15.0-18.0	1.5-1.8	.85 up	.4-.6	31.8-47.8	.45-.65	35.8-51.7	130-220
A	Square Orthonol	14.2-15.8	1.42-1.58	.94 up	.1-.2	7.9-15.9	.15-.25	11.9-19.9	310-715
H	48 Alloy	11.5-14.0	1.15-1.40	.80-.92	.05-.15	4-12	.08-.15	6.4-12	280-550
D	Square Permalloy 80	6.6-8.2	.66-.82	.80 up	.02-.04	1.6-3.2	.022-.044	1.75-3.50	550-1650
R	Round Permalloy 80	6.6-8.2	.66-.82	.45-.75	.008-.02	.64-1.6	.008-.026	.64-2.07	250-715
F	Supermalloy	6.5-8.2	.65-.82	.40-.70	.003-.008	.24-.64	.004-.015	.32-1.19	250-715
S	Supermendur	19-22	1.9-2.2	.90 up	.15-.35	12-27.9	.50-.70	39.8-55.7	85-135
B	Amorphous B	15-16	1.5-1.6	.90 up	.03-.08	2.4-6.4	.04-.1	3.2-7.9	400-900
E	Amorphous E	5-6.5	.5-.65	.90 up	.008-.02	.64-3.2	.01-.025	.79-2	750-2,500
G	Amorphous G	14.5-15.8	1.45-1.58	<.50	Core loss 35 w/lb. @ 10kHz. 1T				

Table 6A48- Trade Names and Gages for Metal Strip Materials

Magnetics Trade Names	Letter Code	Thickness Available	Similar Trade Names
Square Orthonol®	A	.0005" .001" .002" .004" .014"	Orthonik Deltamax Hipernik V 49 Square Mu
Square Permalloy 80‡	D	.0005" .001" .002" .004" .014"	Square Mu 79 Super Square Mu 79 Hy Ra 80 4-79 Permalloy Square Permalloy
Round Permalloy 80‡	R	same as "D" mat'l	Hy Mu 80 Mo-permalloy
48 Alloy	H	.002" .004" .014"	Carpenter 49 Allegheny 4750 Hipernik 49 Alloy
Supermalloy	F	.0005" .001" .002" .004"	Supermalloy
Magnesil®	K	.001" .002" .004" .012"	Silectron Microsil Hypersil Supersil
Supermendur	S	.002" .004"	Supermendur
Amorphous B	B	.001"	METGLAS® Alloy 2605SC
Amorphous G	G	.001"	METGLAS® Alloy 2605S-3
Amorphous E	E	.001"	METGLAS® Alloy 2714A

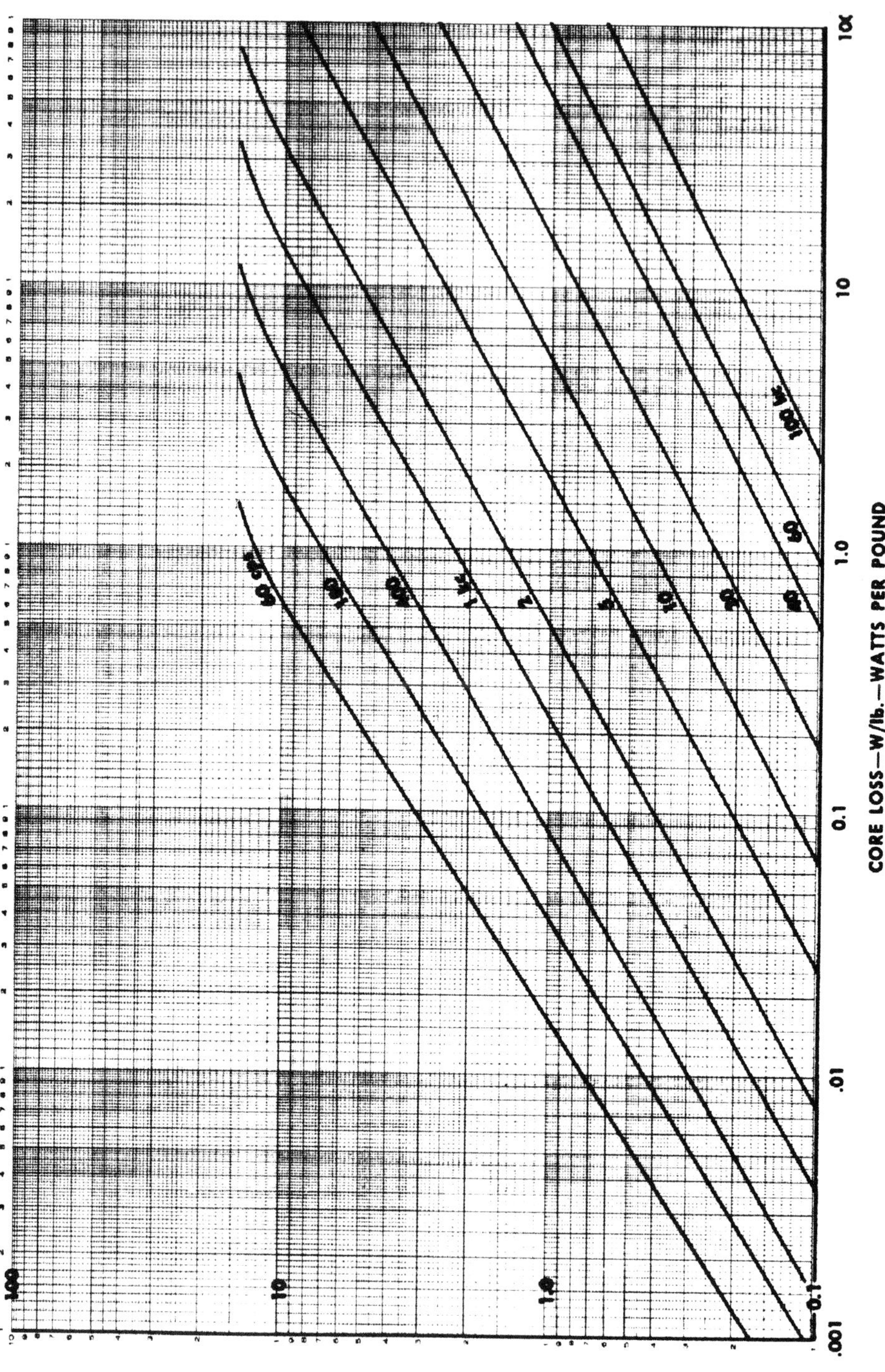

Figure 6A10 - Core Loss Curve vs Frequency for 2 mil Silectron SiFe- Arnold

CORE LOSS CURVES

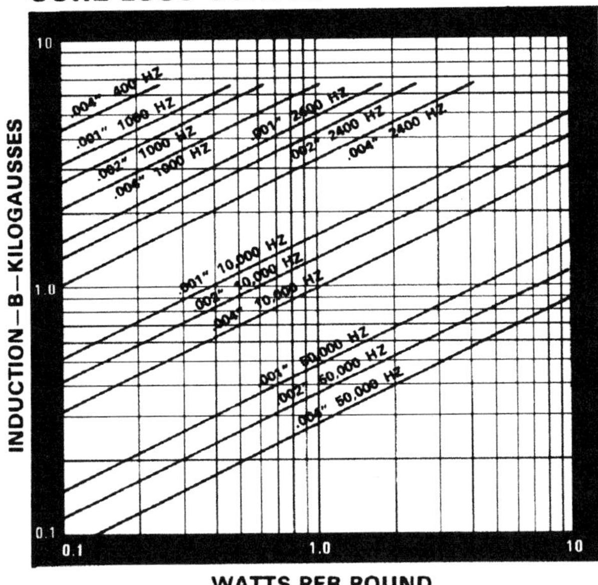

Figure 6A11 - Core Loss Curve for 4-79 Permalloy-Arnold

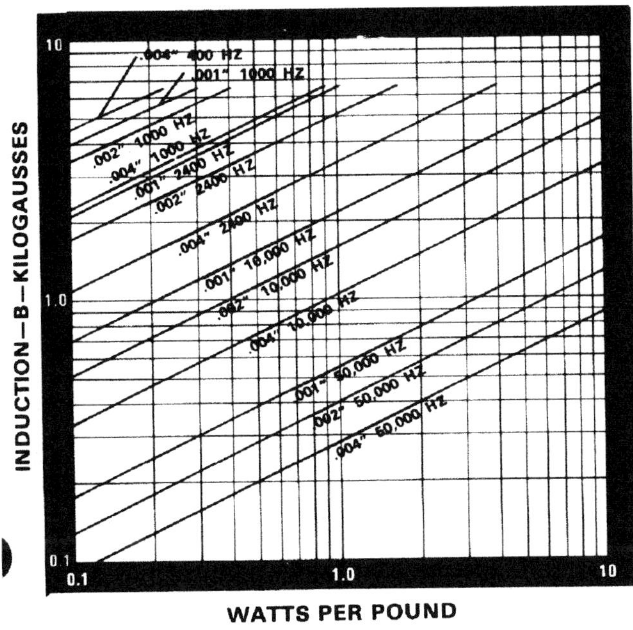

Figure 6A12- Core Loss Curve for Supermalloy-Arnold

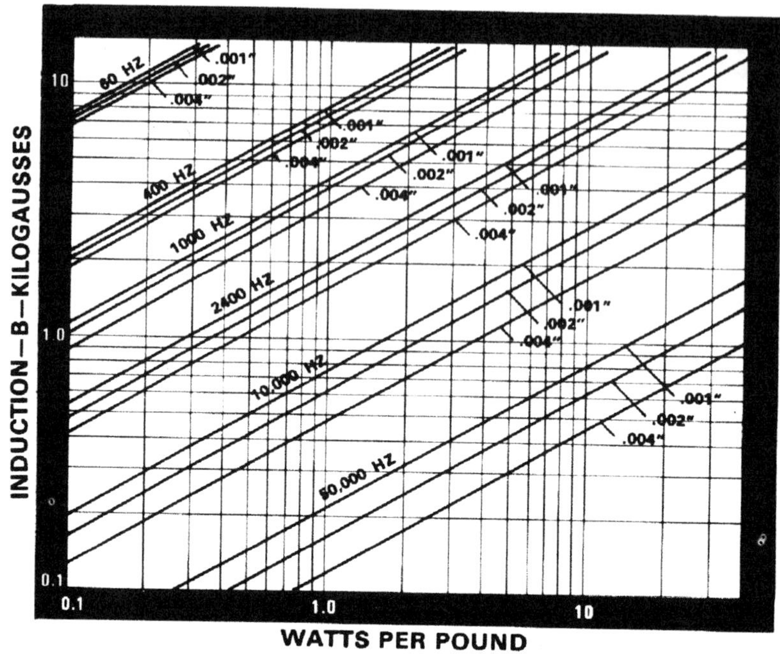

Figure 6A13 - Core Loss Deltamax 50-50 NiFe-Arnold

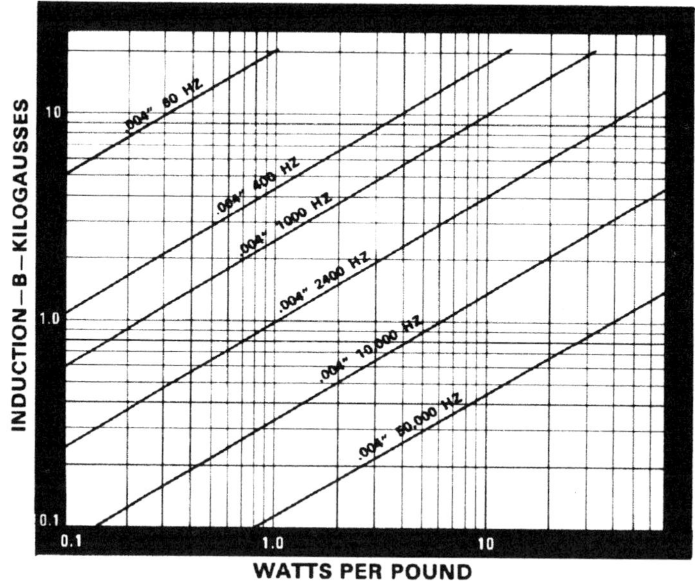

Figure 6A14- Core Loss- Supermendur-Arnold

Table 6A48- Properties of Toshiba Amorphous Alloy

Characteristics \ Core material		Toshiba amorphous alloy	Square permalloy (80%Ni/Fe)
Saturation flux density	kG	7.5	8.0
Density	g/cm^3	8.0	8.7
Curie temperature	°C	320	450
Crystallization temperature	°C	540	-
Resistivity	Ωcm	120×10^{-6}	60×10^{-6}

Characteristic \ Core material		Toshiba amorphous alloy saturable core	Square permalloy (80 Ni) 25um
Coercive force (Oe)	20 kHz	0.07	0.23
	50 kHz	0.13	0.45
	100 kHz	0.21	-
Rectangular ratio (Br/B_1)	20 kHz	0.93	0.94
	50 kHz	0.95	0.95
	100 kHz	0.97	-

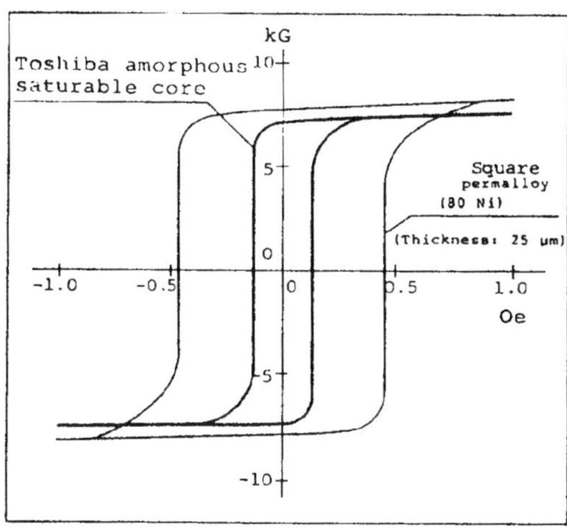

Figure 6A15- Hysteresis Loop for Toshiba Amorphous Alloy

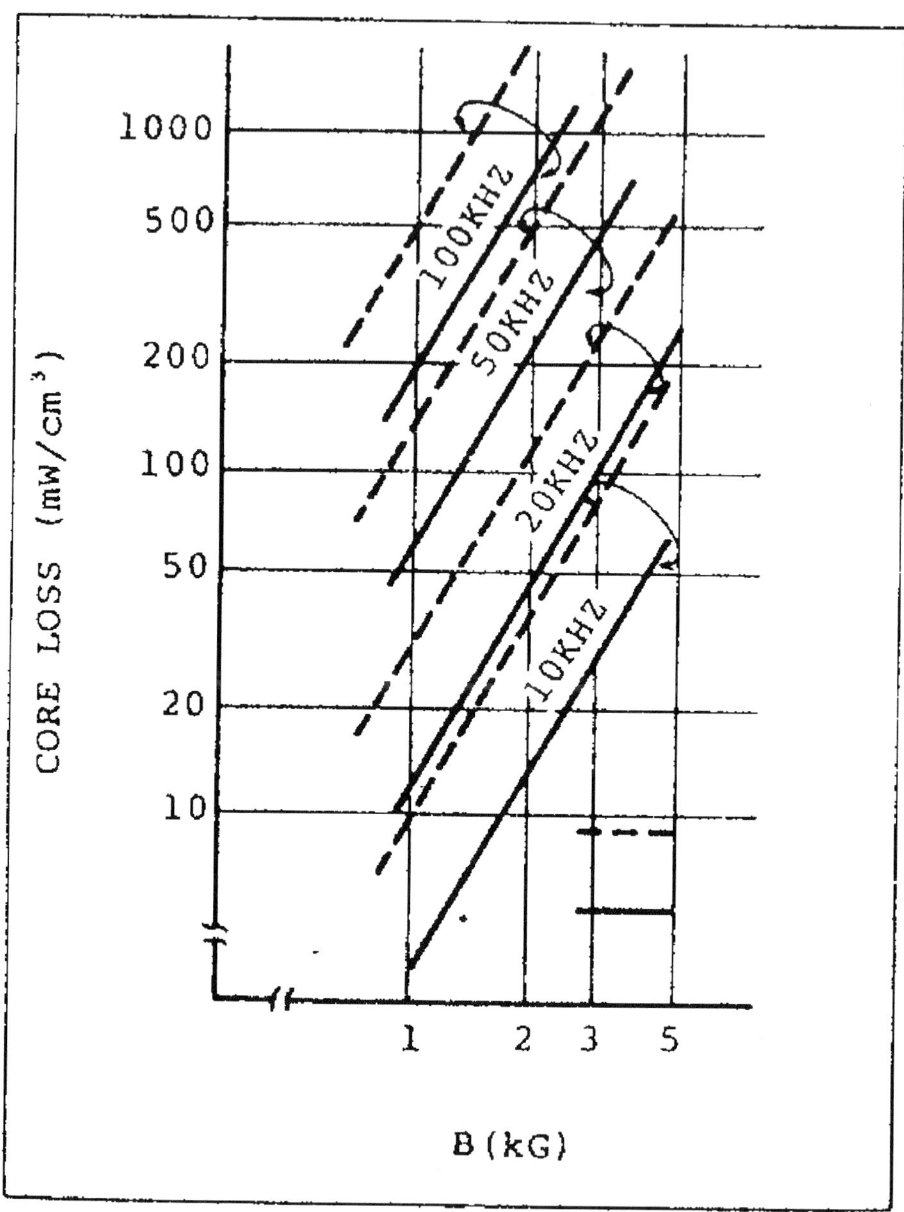

Figure 6A16- Core Loss for Toshiba Amorphous Alloy

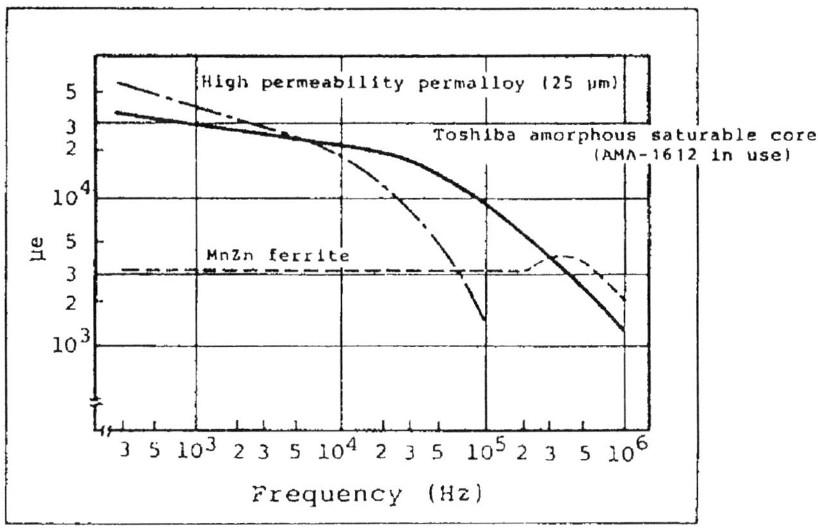

Figure 6A17- Permeability vs Frequency for Toshiba Amorphous Alloy

- High Saturation flux density: 1.2 Tl
- Low Residual flux: <0.2 Tl
- Low losses: <40 mW / g, 0.1 Tl, 200 kHz
- High permeability: 30,000 - 80,000 μ
- High, stable temperature: 120° C
- Outstanding mechanical stability
- Wide operating frequency to 1MHz

APPLICATIONS

- Power transformers for use in AC-DC and DC-DC converters — Especially Forward and Push-Pull Types (because of the very low B_r).
- Current Sense transformers
- Common Mode Inductors for use in switched-mode power Supplies, frequency converters, and UPS units

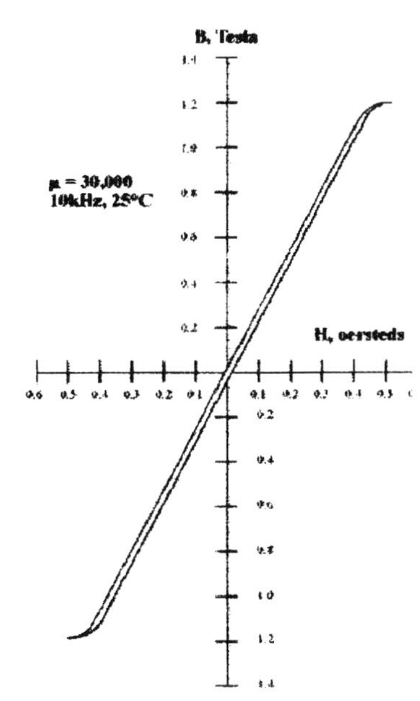

Figure 6A18- Properties of Coremaster E-2000Q Nanocrystalline Cores

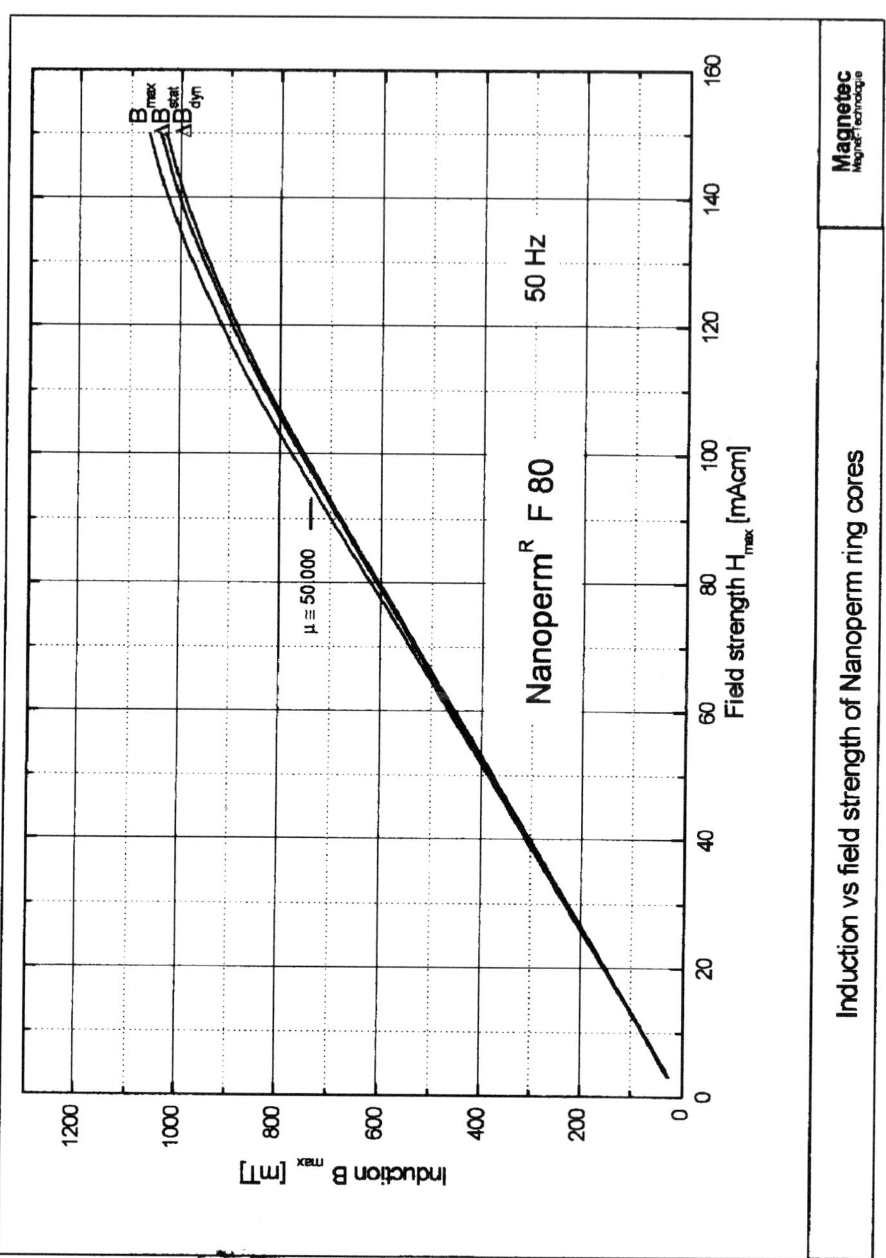

Figure 6A19 -Induction vs Field Strength for Nanpperm®, a Nanocrystalline Alloy

Nanoperm® is a registered trademark of Magnetec GmbH

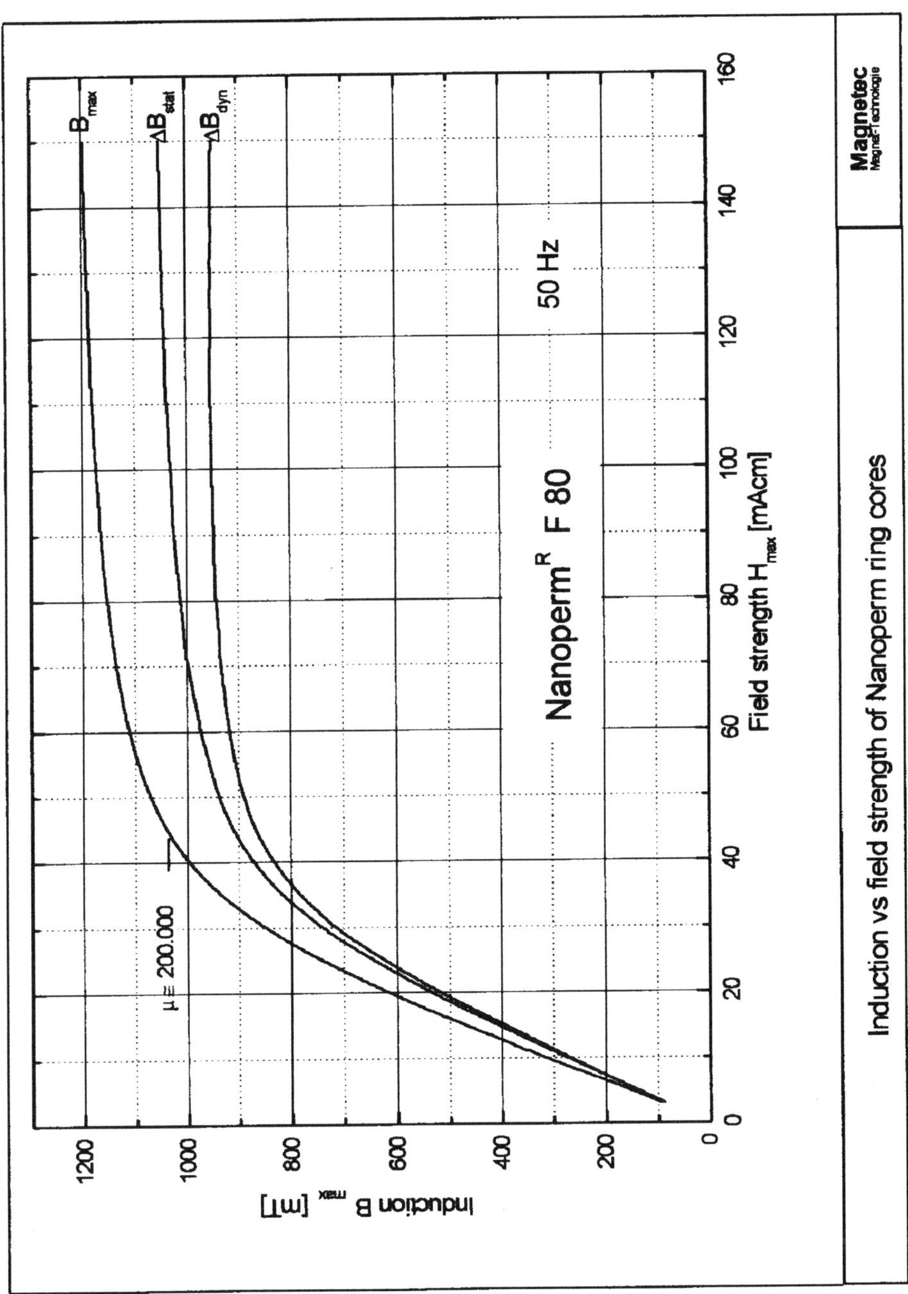

Figure 6A20-Induction vs H -Nanoperm® Nanocrystalline Alloy at 200,000 Perm

Table 6A49- Properties of a Nanoperm® Nanocrystalline Alloy

saturation flux density B_{sat}	1,2	T
saturation magnetostriction λ_s	< 0,5	ppm
spec. electr. resistivity σ	115	$\mu\Omega$cm
density γ	7,35	g/cm3
Curie temperature T_c	600	°C
max. operational temperature T_{max}	120	°C
core losses (0.3T/100kHz,sine) P_v	< 110	W/kg
tape thickness d	17 u. 23	μm
grain size (typ.)	10	nm
permeability levels μ	20.000 - 200.000	
alloy composition	$Fe_{73,5}\ Cu_1\ Nb_3\ Si_{15,5}\ B_7$	

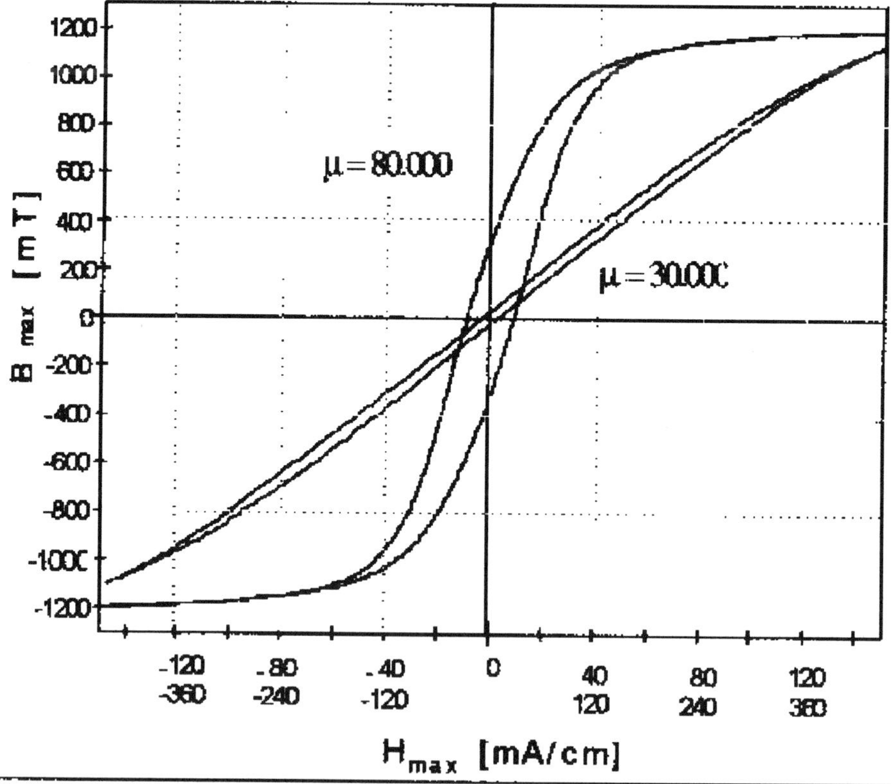

Figure 6A21- Hysteresis Loops of a two Annealed Nanoperm® Nancrystalline Alloys

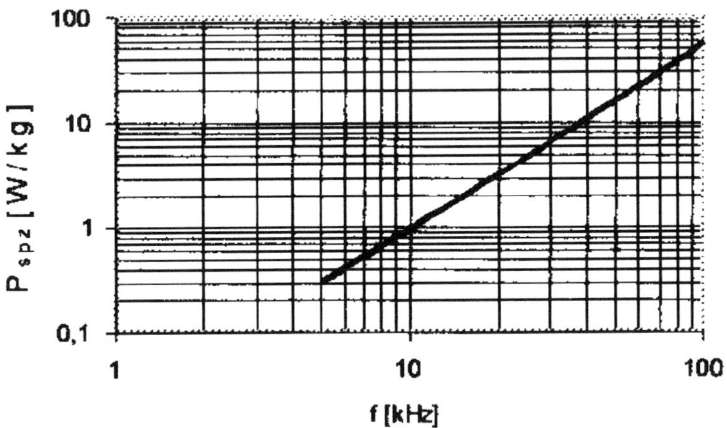

Figure 6A22- Core Loss vs Frequency for a Nanoperm® Nanocrystalline Alloy

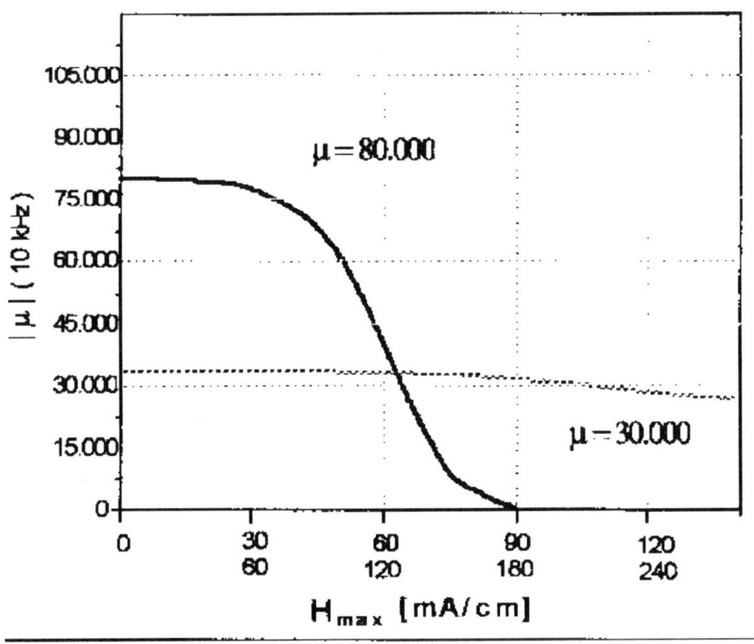

Figure 6A23- Permeability vs DC Bias for a Nanoperm® Nanocrystalline Alloy

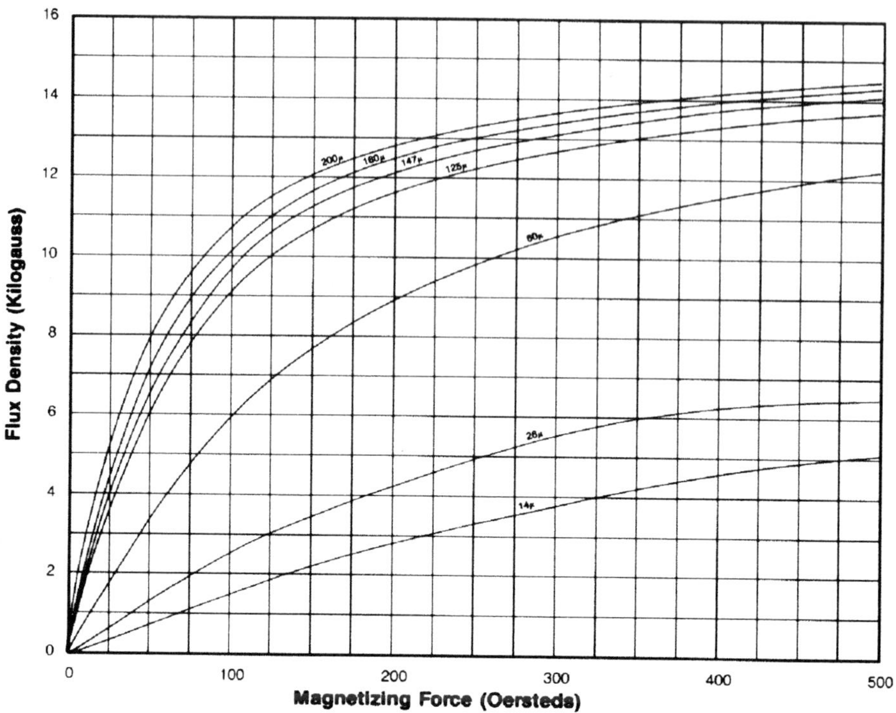

Figure 6A24- Magnetization Curve-Hi Flux -Arnold-Permeabilities from top Curve to bottom are 200, 160,147, 125,60, 26 and 14

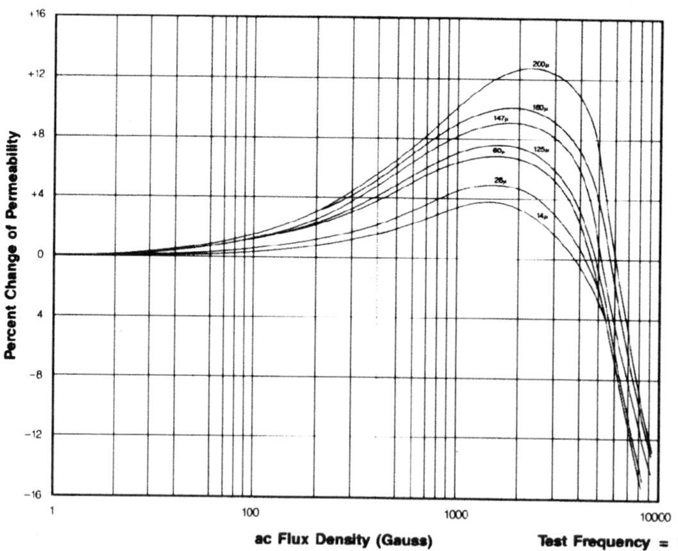

Figure 6A25- Permeability vs Flux Density Hi-Flux Arnold- Permeabilities from top Curve to bottom are 200, 160,147, 125,60, 26 and 14

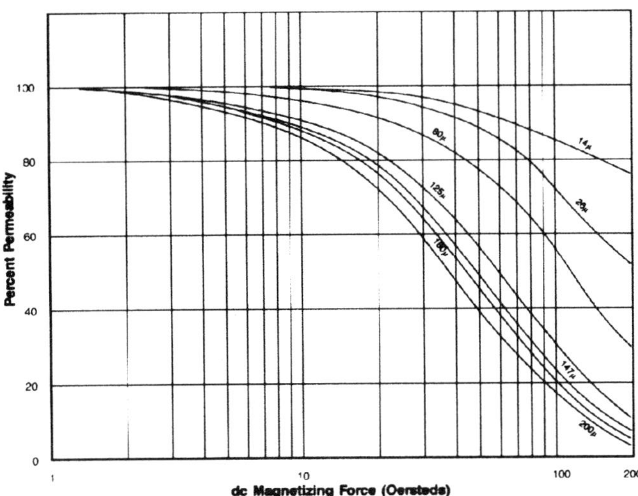

Figure 6A26- Permeability vs DC Bias Hi-Flux Cores-Arnold-- Permeabilities from bottom Curve to top are 200, 160,147, 125,60, 26 and 14

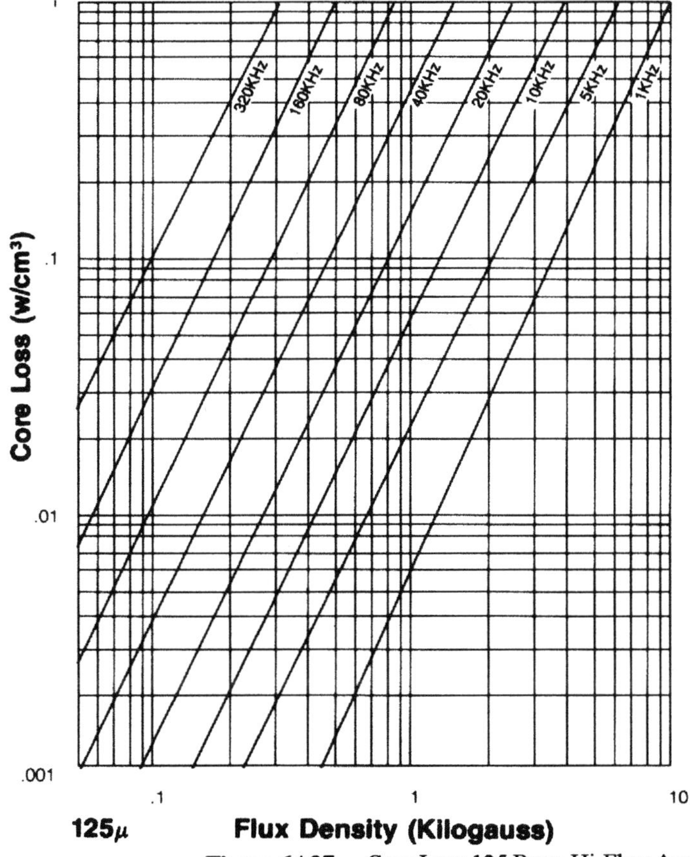

Figure 6A27- Core Loss 125 Perm Hi-Flux Arnold

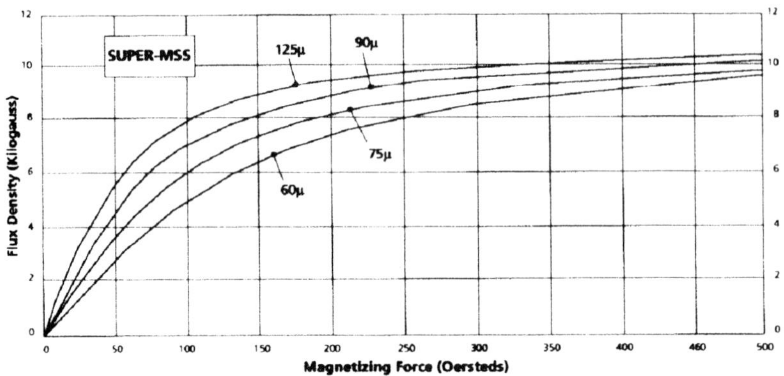

Figure 6A28- Magnetization Curves MSS Arnold

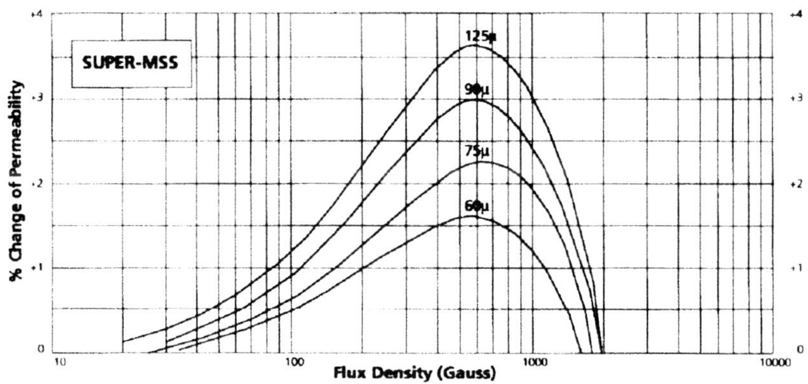

Figure 6A29 - Permeability vs Flux Density MSS-Arnold

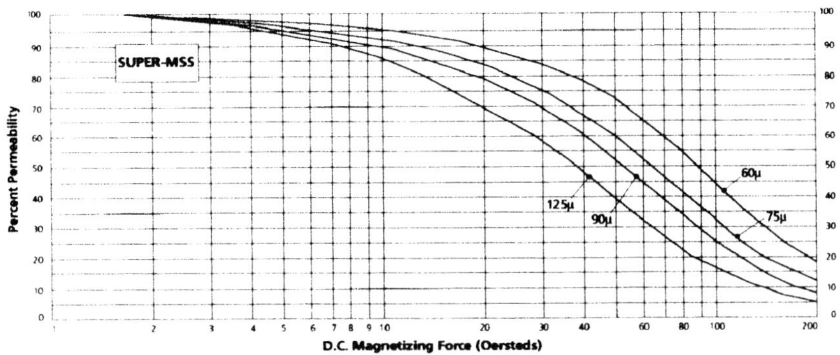

Figure 6A30- Permeability vs DC Bias MSS Arnold

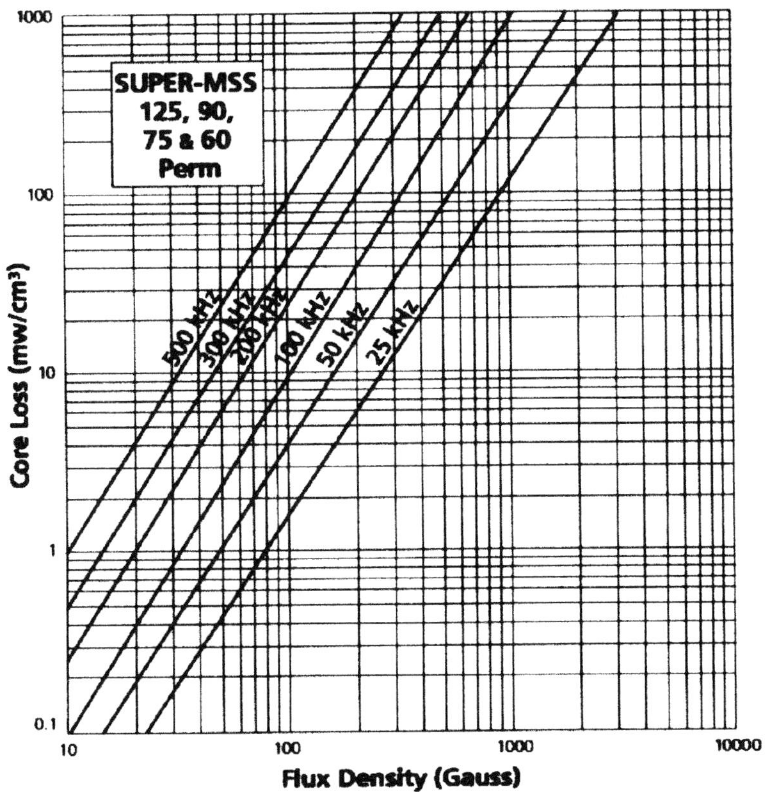

Figure 6A31- Core Loss MSS Arnold

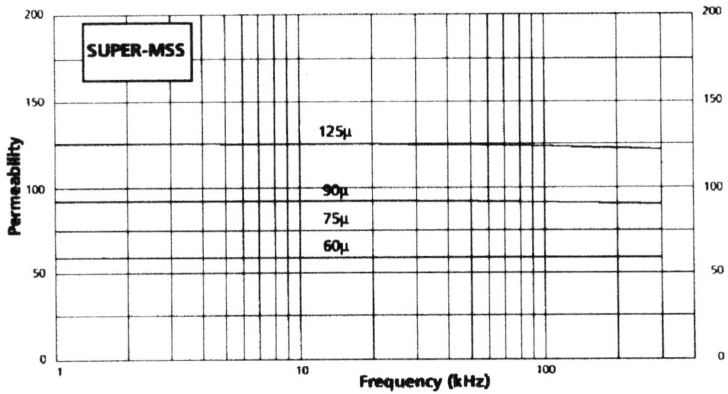

Figure 6A32- Permeability vs Frequency MSS Arnold

Table 6A50- Properties of Vacuumschmelze Amorphous Materials

Typical properties

Alloy		6025	6070	6030	6150
Main transition metal besides metalloids		Co	Co	Co	Co
Saturation flux density	Bs[T]	0,55	0,62	0,82	1,0
Curie temperature	[°C]	210	240	365	485
Saturation magnetostriction 10e-6		<0,2	<0,2	<0,2	<0,2

Measured values on strip-wound cores with Z-loop		6025Z
DC coercivity	[mA/cm]	3
Remanence ratio		0,9
Core losses (100kHz, 0,3T)	[W/kg]	120

Measured values on strip-wound cores with F-loop		6025F	6070F	6030F	6150F
Effective unipolar flux density swing	[T]	0,4	0,5	0,7	0,9
Max. pulse permeability		70.000 - 100.000	25.000 - 50.000	2.500 - 3.500	1.100 - 1.600
Core losses (100kHz, 0,3T)	[W/kg]	100	100	110	130

The mentioned data are nominal values only

Table 6A51- Properties of TDK Amorphous Materials

MATERIAL CHARACTERISTICS

Material				NC20	NA11
Initial permeability	μ_i			250	4500±25%
Curie temperature	T_c	°C		400	410
Saturation magnetic flux density [H = 8000A/m]	B_S	mT	25°C	1150	1180
			60°C	1130	1160
			80°C	1110	1140
			100°C	1090	1120
			120°C	1070	1100
Remanent flux density	B_r	mT	25°C	150	80
			60°C	160	85
			80°C	160	94
			100°C	170	100
			120°C	170	106
Coercive force	H_c	A/m	25°C	160	10
			60°C	160	9.4
			80°C	160	9.1
			100°C	150	8.9
			120°C	150	8.8
Density	d_b	kg/m³		7.2x10³	7.2x10³
Measured core type				T20x10x12	T31x13x19

• *Measuring conditions; direct-current magnetic characteristics.*

Table 6A52.-Properties of Micrometals Powdered Iron Cores

Material Mix No.	Reference Permeability (μ_o)	Material Density (g/cm^3)	Relative Cost	Color Code
-2	10	5.0	2.7	Red/Clear
-8	35	6.5	5.0	Yellow/Red
-14	14	5.2	3.6	Black/Red
-18	55	6.6	3.4	Green/Red
-26	75	7.0	1.0	Yellow/White
-30*	22	6.0	1.4	Green/Gray
-34*	33	6.2	1.5	Gray/Blue
-35*	33	6.3	1.4	Yellow/Gray
-38	85	7.1	1.1	Gray/Black
-40	60	6.9	1.0	Green/Yellow
-45	100	7.2	2.6	Black/Black
-52	75	7.0	1.2	Green/Blue

30 Material was developed as a lower cost, lower loss alternate to the -28 Material. Similarly, the -34 & -35 Materials were ped to replace the -33 Material. The -28 & -33 Materials are not listed in this catalog but are still available.

Table 6A53.-Core Loss of Micrometals Powdered Iron Cores (mW/cm^2)

Material Mix No.	60 Hz @5000G	1kHz @1500G	10kHz @500G	50kHz @225G	100kHz @140G	500kHz @50G	HDC = 50 Oersteds %μ_o	μeffective
-2	19 **	32**	32**	28	19	12	100	10.0
-8	45	64	59	50	35	28	91	31.9
-14	19 **	32**	32**	29	21	17	100	14.0
-18	48	72	70	63	46	37	74	40.7
-26	32	60	75	89	83	139	51	38.3
-30	37	80	120	149	129	129	91	20.0
-34	29	61	87	100	82	78	84	27.7
-35	33	73	109	137	119	123	84	27.7
-38	31	57	72	99	103	217	51	43.4
-40	29	62	93	130	127	223	62	37.2
-45	26	49	60	69	61	92	46	46.0
-52	30	56	68	72	58	63	59	44.3

** Low frequency core loss is extrapolated from data measured at high frequency

Table 6A54.-Permeabilities-Micrometals Iron Powder Core Applications

MATERIAL APPLICATIONS

Typical Application	-2	-8	-14	-18	-26	-30	-34	-35	-38	-40	-45	-52
Light Dimmer Chokes					X				X	X	X	
60 Hz Differential-mode EMI Line Chokes					X				X	X	X	X
DC Chokes: <50kHz or low Et/N (Buck/Boost)					X	X	X	X	X	X	X	
DC Chokes: ≥50kHz or higher Et/N (Buck/Boost)		X	X	X		X	X	X				X
Power Factor Correction Chokes: <50kHz					X	X	X	X		X		
Power Factor Correction Chokes: ≥50kHz	X	X	X	X		X	X	X				
Resonant Inductors: ≥50kHz	X		X									

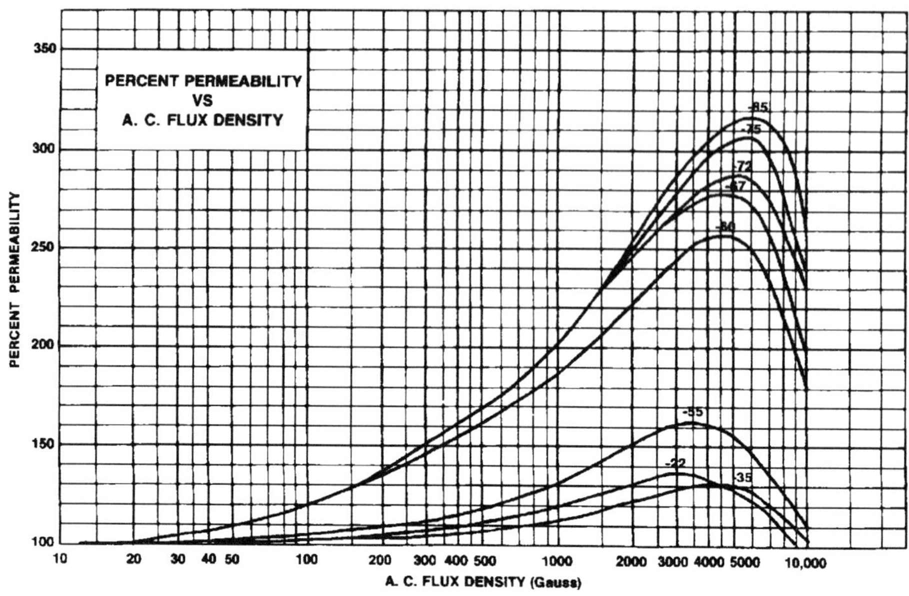

Figure 6A33- Permeability vs ac flux density for several different permeability iron powder cores. Curve numbers are the permeabilities. From Pyroferric

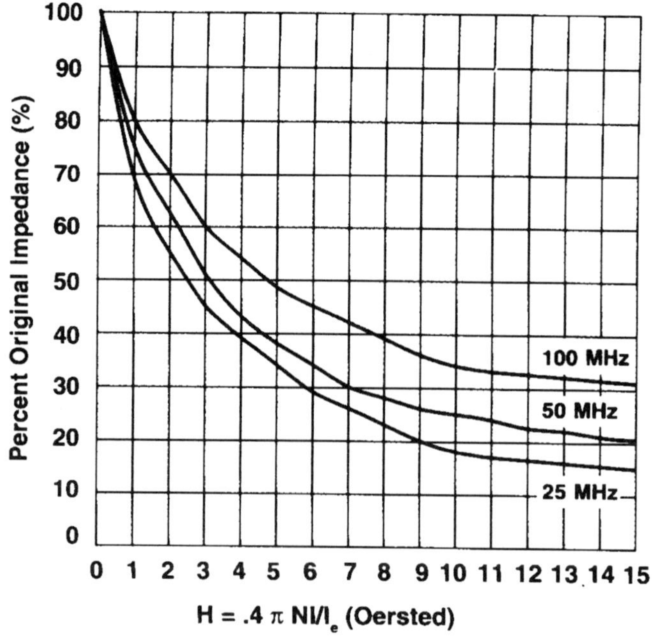

Figure 6A34-Percent reduction in permeability with DC bias for several different permeability iron powder cores. Curve numbers are the perms.From Pyroferric

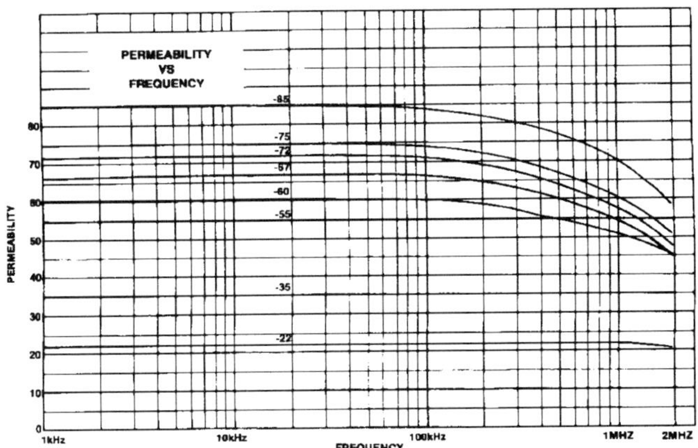

Figure 6A35-Variation of permeability with frequency for several different permeability iron powder cores From Pyroferric

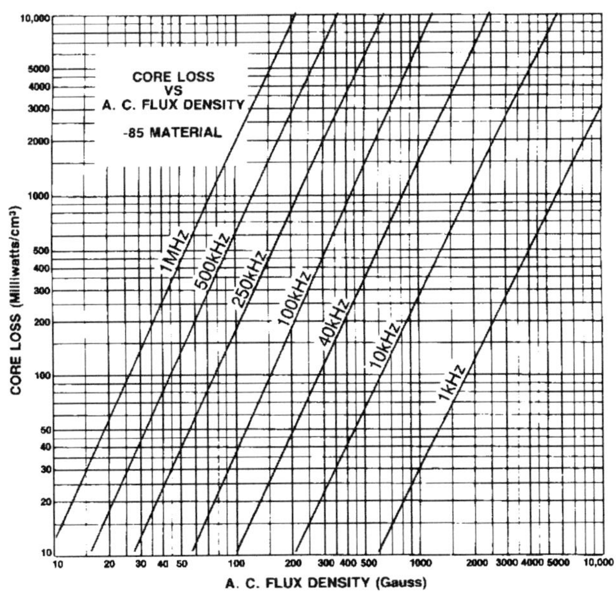

Figure 6A36- Core Losses of a 85 perm iron powder core. From Pyroferric

Table 6A55.-Pyroferric Powdered Iron Permeabilities

Material Suffix	Permeability	Color Code	L Tolerance
-33	33	White/Gray	± 10 %
-35	35	White/Green	± 10 %
60	60	Gray	± 10 %
-67	67	Gray/Green	± 10 %
-75	75	Gray/White	± 10 %
-LL	75	Gray/Blue	± 10 %
-85	85	White	± 10 %

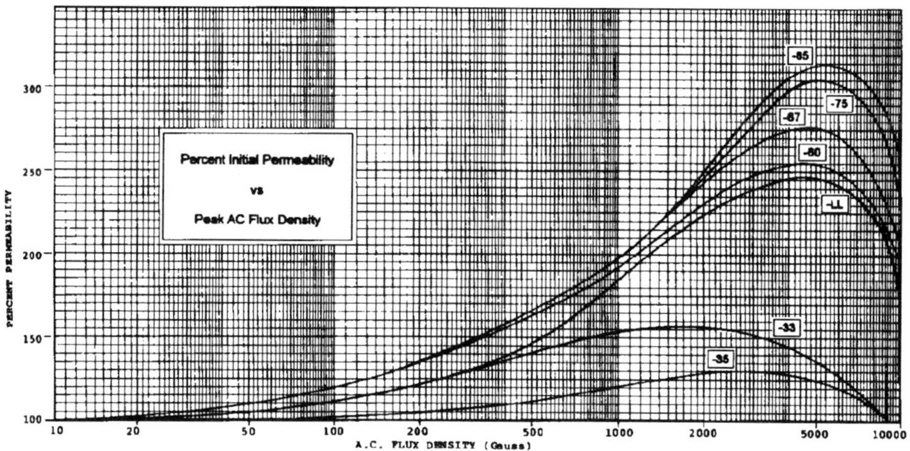

Figure 6A37.-Permeability vs B for Pyroferric Powdered Iron Cores

Summary

This chapter has listed the properties of a great number of materials and cores from the major power magnetic component vendors. The final chapter will be present a compilation of many of the design aids that are available to the engineer who is making the choice of the optimum power magnetic component.

Chapter 7
DESIGN AIDS IN MAGNETIC COMPONENT CHOICE FOR POWER ELECTRONICS

INTRODUCTION

In the early days of power electronics, the design engineer had few design aids in magnetic component choice. As a result, the choice was made by using whatever component happened to be on the shelf and after numerous hit-and miss tries, he finally found the most suitable choice. At that time, both power semiconductor and high frequency power materials were primitive with very little choice of materials or shapes. However, today with the explosion of information through media like the Internet, CD ROM, CAD-CAM, and computers in general, there is a great deal of help that makes the proper choice faster and with greater assurance of success. This chapter will review the various design aids available. In Appendix 7.1, articles of interest on Power electronics are listed. In Appendix 2, the pertinent IEC and ASTM standards on magnetic components and Materials are listed.

7.1- Books on Power Electronics

The following is a compilation of some of the power electronics books listed by EJBloom Associates;

Books on Power Magnetics

1. Transformer and Inductor Design(2nd Edition)- C. McLyman
2. Magnetic Core Selection for Transformers and Inductors (Second Edition)- C. McLyman
3. Applications of Magnetism-J.K. Watson
4. Handbook of Transformer Applications- W. Flanagan
5. Designing Magnetic Components for High frequency DC-DC Converters- C. Mc Lyman
6. Handbook of Modern Ferromagnetic Materials- A. Goldman

Books on Power Electronic Circuits

1. Modern DC-DC Switchmode Power Converter Circuits- Severns and Bloom
2. Switching Power Supply Design(2nd Edition) A. Pressman

3. High Frequency Resonant and Soft Switching Converters-VPEC Staff
4. Power Supply Cookbook(Second Edition) M.Brown
5. Fundsamentals of Power Electronics(2nd Edition)- R. Erickson and D. Maximovic
6. Power-Switching Converters-S.Ang
7. Elements of Power Electronics-P.Krein
8. 1995 VPEC Seminar Proceedings
9. 1996 VPEC Seminar Proceedings
10. 1997 VPEC Seminar Proceedings
11. 1998 VPEC Seminar Proceedings
12. 1995 VPEC Seminar Proceedings
13. 1999 VPEC Seminar Proceedings
14. Introduction to Modern Power Electronics-A. Trzynadlowski
15. 2000 VPEC Seminar Proceedings w/CDROM
16. Advanced Soft-Switching Techniques, Device and Circuit Application-VPEC/CPES Staff

PSMA Industry Standards and Publications

1. The Power Technology Roundup Report (Year 2000)-PSMA
2. Handbook of Standardized Technology for the Power Sources Industry (2nd Edition)-PSMA
3. Low Voltage Study-Workshop Report L. Pechi-PSMA

Circuit Design

1. Switch Mode aPower Conversion-K.Sum
2. SPICE-A Guide to Circuit Simulation & Analysis using Pspice(3rd Edition)-w?IBM Program Disk-P. Tuinenga
3. Modeling, Analysis & Design of PWM Converters-VPEC Staff
4. Pspice Simulation of Power Electronics-R. Ramshaw
5. SMPS Simulation with SPICE 3 (w/Diskette)-S.Sandler
6. Circuit Simulation of Switching Regulators Using SPICE-(w/Diskette)-V. Bello
7. Switch-Mode Power Supply SPICE Cookbook (Includes CDROM)-C.P.Basso

7.2-Power Electronics Magazines

The following magazine are concerned with power electronics;

1. PCIM Power Electronics
   Intertec Publishing Co.
   9800 Metalf Ave.
   Overland Park KS 66212

2. Switching Power Magazine
   Ridley Engineering Co. Ltd.
   1575 Old Alabama Road Suite 207-147
   Roswell, GA 30076
3. Magnews
   UK Magnetics Society
   Berkshire Business Centre
   Post Office lane
   Wantage, Oxon, OX12H, UK
4. IEEE Transactions on Magnetics
   IEEE
   445 Hoes Lane
   P.O. Box 1331
   Piscataway, NJ 08855-1331
5. EDN
   Cahners Publishing Co.
   Cambridge, MA
6. Power Pulse
   Darnell Group
   1159 B. Pomoma Road
   Corona, CA 92882-6926
7. Power Quality

7.3-Power Electronics Organizations

There are a number of government organizations, societies and private groups involved in power electronic. The are;

1. IEEE Power Electronics Society
   IEEE Power Electronics Society
   Robert Meyers, Admin.
   799 N. Beverly Glen
   Los Angeles CA 90077
2. Power Electronics & Electronics Machinery Research Center
   Oak Ridge National Laboratory
   U.S. Dept.of Energy
   P.O. Box 2009
   Oak Ridge TN 37831-2009
3. IEEE
   445 Hoes Lane
   P.O. Box 1331
   Piscataway, NJ 08855-1331
3. Power Sources Manufacturers Association
   P.O. Box 418

Mendham, NJ 07945-0418
4. Electric Power Institute
5. National Science Foundation
6. PCIM

7.4 Power Electronic Web Sites

There are many Web sites related to power electronics. We will try to list all we now;

1. pesc.org
2. orln.gov/etd/peemc/PEEMCHome.htm
3. smpstech.com
4. psma.com
5. efore.fi/
6. jobsearch.monster.com
7. greshampower.com
8. tycoelectronics.com
9. spangpower.com
10. ardem.com (R.D. Middlebrook)
11. pels.org
12. home.aol.com/DrVGB (V. Bello)
13. venableind.com
14. darnell.com
15. cpes.vt.edu
16. kgmagnetics.com

7.5 Power Electronic Conferences

The Power Electronics Society Sponsors or co-sponsors several conferences. They are

1. Power Electronics Specialist Conference
2. Applied Power Electronics Conference
3. International Telecommunications Energy Conferenc

There are other Conferences related to the Magnetics community;

1. Intermag Conference
2. Magnetism and Magnetic Materials Conference(MMM)
3. International Conference on Ferrites
4. Soft Magnetic Materials Conference
5. Intertech Conferences on Magnetic Materials

7.6-Power Electronics at Universities and Research Labs

Listed below are the universities and research labs that are doing major project in power electronics;

1. Colorado Power Electronics Center (CoPEC)
2. Caltech Power Electronics Group
3. School of Electrical Engineering, University of Belgrade, Yugoslavia
4. UC Berkley
5. UC Irvine Power Electronics Laboratory
6. Virginia Power Electronics Laboratory(VPEC)
7. University of Wisconsin-Madison(WEMPEC)
8. Georgia Tech Electric Power Research
9. North Carolina State Electric Power Research Center
10. Surrey University Electrostatic Systems and Power Electronics Research Group
11. Texas A&M University
12. Northeastern University
13. University of Central Florida
14. JPL Power Electronics Group
15. University of Waterloo

7.7- Web Tutorials

Listed below are some of the Internet tutorial sites;

1. Switching Power Supply Design-Jerrold Foutz at smpstech.com
2. Interactive Power Electronics OnlineText- ee.uts.edu.au/~venkat/
3. SPICE-A Brief Overview-seas.upenn.edu/~jan.spice.overview.htm
4. Power Factor; Dissipating the Myths- Microconsultants.com/tips/pwrfact /pfartil

7.8 Power Electronics Software

There are many power electronics software programs. Listed below are the ones listed on the ejbloom associates website

1. Power Witts™ Forward Converter Simulation
2. Pspice Electronic Simulation-N.Mohan
3. Computer Aided DesignforIndictors and Transformers-KG Magnetics
4. Magnetic Core Data for Converters-KG Magnetics
5. Flyback Converter Magnetic Design
6. Specialty I-Magnetic Design-DC-DC Converters-KGMagnetics
7. Computer-Aided Inductor and Transformer Design-KG Magnetics
8. Specialy II-Common-Mode Chokes-KG Magnetics

9. Titan Systems-KG Magnetics

7.9-Power Electronic Short Courses

EJBloom Associates runs frequent courses on Power Electronics. For information log on to ejbloom.com or e-mail to ejbloom@compuserve .com. The address is;

EJ Bloom Associates
115 Duran Drive
San Rafael, CA 94903

7.10- Component Vendors CDROM or Diskettes

Many suppliers offer product information and design on CDROM disks or IBM diskettes. Below is a list of some that are available

1. FDK
2. Ferroxcube
3. Tokin
4. Metglas^R (Honeywell)
5. Epcos
6. Magnetics
7. Kaschke

7.11-INTERNET WEB SITES

Just about all the manufacturers of ferrites have Web sites on the Internet. In most cases, the core data can be down loaded and printed by the user. Where large catalogs are involved, the use of the Acrobat reader is needed but the vendor can often download this as well. Some of the vendors that maintain Web sites are;

1. Magnetics
2. Philips
3. Siemens
4. Fair-Rite Products
5. Steward
6. MMG North America
7. Micrometals

7.12-Magnetic Component Standards

In Appendix 7.2 below are the appropriate IEC and ASTM Standards for Magnetic Components and Magnetic Materials

Appendix 7.1

Recent Power Electronic Articles on Specific Power Electronic Subjects

Planar Magnetics

1. Designing Planar Magnetics, J. Martinos, PCIM, August 2000, 102
2. Planar Magnetics, Philips Components, PC1012, 2/2/94
3. Planar Transformers Dispense with Bulkiness, D. Maliniak, Electronic Design, Feb. 1993, 34
4. Design Techniques for Planar Windings with Low Resistance, HX. Huang, D.T. Ngo and G. Bloom, IEEE Appl. Power Elect. Conf. (1995)
5. Planar Magnetics Lower Profile Improve Converter Efficiency, A. Estrov and I. Scott PCIM, May 1989, 16
6. Planar Magnetics Simplifies SMPS Design and Production, E. Brown, PCIM, July 1992, 46
7. Design of a HF Planar Power Transformer in Multilayer Technology, D. van der Linde C.A.N.Boon and J. B. Klaasens, IEEE Trans. Ind. Electr., 38, (1991) 137
8. Effects of Air Gaps on Winding Loss in HF Planar Magnetics, K.D.T. Ngo and M.H. Kuo, PESC '88 Record april 1988
9. A Comparative Study of HF, Low Profile Planar Transformwer Technologies, A.W. Lofti, N. Dai, G. Skutt, W. Tabisz and F.C. Lee, Va. Power Electr. Center EPE, 1993
10. Planar High Density Design for Automotive and Telecommunications Applications, High Frequency Magnetic Materials Workshop '99,1999 Santa Clara, CA Organized by Gorham-Intertech, Portland Maine,
10. Planar Ferrites, M. A. Swihart, PCIM, July 1999, 12
11. Planar Transformers Dissipate Up to 150 kW with Copper Trace Windings, G.T. Tarzwell, PCIM, March 2000, 58

Integrated Magnetics

1. Issues in Flat Integrated Magnetics Design, E. Santi and S. Cuk, 1996 Power Electronics Spec. Conf .Proc. Vol.
2. Integrated PC Board Transformers Improve PWM Converter Performance, B. E. Mohandes, PCIM, July 1994, 8
3. Integrated Magnetics Design with Flat Low Profile Cores, S. Cuk, F.
4. Stavaovic and E. Santi, Power Elecronics Group,Cal. Inst. Of Technology

Surface Mount Design

1. Manufacturing Advances Increase Availability of Surface Mount Magnetics, B. Tscosik and W. Dangler PCIM, July 1994, 60
2. Surface Mount Transformers, Surface Mount Technology, Feb 1993, 27
3. Isolated Innovation Marks Movement Towards Miniature Magnetics, R.A. Quinnell, EDN, July 1994, 59

Magnetic Core Materials

1. Core Materials, D.J.Nicol, PCIM, July 1999,58

2. Power Supply Magnetics- Part II-Selecting a Core Material, D.E. Pauly, PCIM, Feb. 1996,36
3. Data Sheets Provide Clues for Optimizing Selection of SMPS Ferrite Core Material, L.M. Bosley, PCIM, July 1993, 13
4. High Frequency Magnetic Materials and Markets, L.M. Bosley and T. Holmes, High Frequency Magnetic Materials, '99, Santa Clara CA. Organized by Gorham-Intertech, Portland MN.
5. Soft Ferrites,Potential-Further Development, R. Dreyer and L. Michalowsky, ibid
6. High Frequency Applications of Amorphous Metal, R. Hasagawa, ibid
7. Ferrite "Boost" Material Improves Inductor Characteristics under DC Bias Conditions, G. Orenchak, PCIM, Nov. 1999, 48
8. Common-Mode Inductors for EMI Filters Require Careful Attention to Core Material Selection, R. West, PCIM, July 1995, 52
9. Inductor Core Material Operates Efficiently at High Frequencies, W.A. Martin , PCIM, July 1992, 10

Design

1. Core Selection and Winding Design for HF Magnetics, R. Severns, , High Frequency Magnetic Materials, '99, Santa Clara CA. Organized by Gorham Intertech, Portland MN.
2. Designing Flyback Transformers Part I.-Design Basics, R. Ruble and R. Clarke, PCIM, Jan. 1994, 43
3. Designing Flyback Transformers Part II-48 V. Dual Outpt DC-DC Converter., R. Ruble and R. Clarke, PCIM, April 1994, 23
4. CAD Software and Experience Cut High Power Transformer Design Time and Cost, PCIM, April 1992,50
5. Advances in Magnetic Modeling Using Finite Element Analysis, M. Christini, , High Frequency Magnetic Materials, '99, Santa Clara CA. Organized by Gorham Intertech, Portland MN.
6. Redesigned Input Filter Limits DC-DC Converter Inrush Current, B. Bell, PCIM, March 2000, 66
7. DC-DC Converter Power Density Design Issues, W. Leitner, PCIM, Sept. 1990, 22
8. Power Ferrite Loss Formulas for Transformer Design, S. Mulder, PCIM, July 1995, 22
9. System Integration of a DC-DC Converter Requires a Look at All Angles, S. Davis, PCIM, Oct. 2000, 88
10. Power Factor Basics, M. Amato, PCIM, Oct. 1995, 10
11. Expert System/Fuzzy Logic Select Core Geometry for HF Power Transformers, Part I and II, R. K. Dhawan, P. Davis and R. Naik, PCIM, April and May 1995, 44 and 34
12. Understand Data Book Parameters When Selecting DC-DC Converter Inductors, L. Crane, PCIM, Oct. 2000, 54

13. Keep Core Geometry In Mind When Designing Transformers, C.R. Wild, PCIM, July 2000, 32
14. DC-Bias Behavior Calculation for Uniform and Non-Uniform Cross-Section Ferrite Cores,, D. Lange and S. Ahne, PCIM, March 2000, 12
15. Transformer Bobbin and Core Selection Involves Interdisciplinary Design and Cost Issues, J. Casmero and R. Barden, PCIM, Oct. 2000, 20

Applications

1. Power Supply Basics-Voltage Regulators, R.J. Shah, PCIM, April 1996, 20
2. Optimizing Flyback Techologies for Portable AC/DC Adapter/Charger Applications-Part I:Adapter/Charger Requirements, L.Huber and M.M. Jovanovic, PCIM,August 1996, 68
3. Low-Cost Battery Charger Transformer Design Require Performance and Safety Considerations, R.M. Haas, PCIM, June 1994, 46
4. EMI Design Factors for Medical-Grade Switching Power Supplies, R. Hood and J. Belna, PCIM, Oct. 1992, 32
5. New Magnetic Circuit Design Yields High orce, Long Stroke Linear Actuators, B. Bartosh, PCIM, Oct. 1991,38
6. Telcom Power, D.L. Cooper, PCIM, April 1999,38
7. Automotive Power, J.M. Miller, P.R. Nicastri and S. Zarei, PCIM, Feb. 1999, 44
8. Low-Cost UPS Guards Against Data Loss, S. Davis, PCIM, July 1992, 32
9. Mag Amps, J. Goodin, PCIM, July 1998, 30
10. Low Profile Inductors, Semiconductors and Capacitors Make PCMCIACard DC-DC Converters Possible, B. Tschosik,PCIM, June 1996, 9

Flexible Circuits

1. Magnetics Component, M. San Roman and D. Longden,PCIM, July, 1999, 67
2. 1-MHz Resonant Converter Power Transformer Is Small, Efficient, Economical,A. Estrov, PCIM, Aug. 1986, 14
3. Converters Pull for Flat Magnetism, Y. B. Salih, Electr. Eng. Times, Aug. 1992, 52
4. Building Magnetics with Flexible Circuits, V. Gregory, Powertechnics Mag. (1989), 16

EMI Applications

1. Selecting EMI Suppression Ferrites Requires Attention to Complex Permeability C.U. Parker, PCIM, Feb. 2000, 62
2. Transients vs Electronic Circuits, Part IV, Ferrites for EMI Suppression, PCIM, July 1996, 76
3. Power Analysis for IEC 1000-3-2/3- Q&A. T. Mahr, PCIM, Feb. 2000,82

Appendix 7.2- IEC and ASTM Standards

133	(1985)	Dimensions of pot-cores made of magnetic oxides and associated parts. (Third Edition).
205	(1966)	Calculation of the effective parameters of magnetic piece parts. Amendment No. 1 (1976). Amendment No. 2 (1981).
205A	(1968)	First supplement.
205B	(1974)	Second supplement.
220	(1966)	Dimensions of tubes, pins, and rods of ferromagnetic oxides.
221	(1966)	Dimensions of screw cores made of ferromagnetic oxides. Amendment No. 2 (1976).
221A	(1972)	First supplement.
223	(1966)	Dimensions of aerial rods and slabs of ferromagnetic oxides.
223A	(1972)	First supplement.
223B	(1977)	Second supplement.
226	(1967)	Dimensions of cross cores (X-cores) made of ferromagnetic oxides and associated parts. Amendment No. 1 (1982).
226A	(1970)	First supplement.
367:	–	Cores for inductors and transformers for telecommunications.
367-1	(1982)	Part 1: Measuring methods. (Second Edition). Amendment No. 1 (1984). Amendment No. 2 (1992).
367-2	(1974)	Part 2: Guides for the drafting of performance specifications. Amendment No. 1 (1983).
367-2A	(1976)	First supplement.
401	(1993)	Ferrite materials - Guide on the format of data appearing in manufacturers' catalogues of transformer and inductor cores. (Second Edition).
424	(1973)	Guide to the specification of limits for physical imperfections of parts made from magnetic oxides.
431	(1993)	Dimensions of square cores (RM cores) made of magnetic oxides and associated parts. (Second edition). Amendment No. 1 (1995).
492	(1974)	Measuring methods for aerial rods.

Table 23.2(Continued)

525	(1976)	Dimensions of toroids made of magnetic oxides or iron powder. Amendment No. 1 (1980).
647	(1979)	Dimensions for magnetic oxide cores intended for use in power supplies (EC cores).
701	(1981)	Axial lead cores made of magnetic oxides or iron powder.
723:	—	Inductor and transformer cores for telecommunications.
723-1	(1982)	Part 1: Generic specification.
723-2	(1983)	Part 2: Sectional specification: Magnetic oxide cores for inductor applications.
723-2-1	(1983)	Part 2: Blank detail specification: Magnetic oxide cores for inductor applications. Assessment level A.
723-3	(1985)	Part 3: Sectional specification: Magnetic oxide cores for broadband transformers.
723-3-1	(1985)	Part 3: Blank detail specification: Magnetic oxide cores for broadband transformers. Assessment levels A and B.
723-4	(1987)	Part 4: Sectional specification: Magnetic oxide cores for transformers and chokes for power applications.
723-4-1	(1987)	Part 4: Blank detail specification: Magnetic oxide cores for transformers and chokes for power applications - Assessment level A.
723-5	(1993)	Part 5: Sectional specification: Adjusters used with magnetic oxide cores for use in adjustable inductors and transformers.
723-5-1	(1993)	Part 5: Sectional specification: Adjusters used with magnetic oxide cores for use in adjustable inductors and transformers. Section 1: Blank detail specification - Assessment level A.
732	(1982)	Measuring methods for cylinder cores, tube cores and screw cores of magnetic oxides.
1007	(1994)	Transformers and inductors for use in electronic and telecommunication equipment -Measuring methods and test procedures. (Second Edition).
1185	(1992)	Magnetic oxide cores (ETD-cores) intended for use in power supply applications - Dimensions. Amendment No. 1 (1995).
1246	(1994)	Magnetic oxide cores (E-cores) of rectangular cross-section and associated parts - Dimensions.

IEC Standards for Metallic Materials

Under the Responsibility of IEC/TC68 *MAGNETIC ALLOYS AND STEELS*

IEC 60404-1: 1979	Magnetic materials. Part 1: Classification. (A new version will be published in 1999)
IEC 60404-2: 1996	Magnetic materials. Part 2: Methods of measurement of the magnetic properties of electrical steel sheet and strip by means of an Epstein frame.
IEC 60404-3: 1992	Magnetic materials. Part 3: Methods of measurement of the magnetic properties of magnetic sheet and strip by means of a single sheet tester.
IEC 60404-4: 1995	Magnetic materials. Part 4: Methods of measurement of the d.c. magnetic properties of iron and steel.
IEC 60404-5: 1993	Magnetic materials. Part 5: Permanent magnet (magnetically hard) materials - Methods of measurement of magnetic properties.
IEC 60404-6: 1986	Magnetic materials. Part 6: Methods of measurement of the magnetic properties of isotropic nikel-iron soft magnetic alloys, types E1, E3 and E4.
IEC 60404-7: 1982	Magnetic materials. Part 7: Methods of measurement of the coercivity of magnetic materials in an open magnetic circuit.
IEC 60404-8-1: 1986	Magnetic materials. Part 8: Specifications for individual materials. Section One - Standard specifications for magnetically hard materials. Amendment 1: 1991 Amendment 2: 1992, (A new version will be published in 1999)
IEC 60404-8-2: 1985	Magnetic materials. Part 8: Specifications for individual materials. Section Two - Specification for cold-rolled magnetic alloyed steel strip delivered in the semi-processed state. (A new version will be published in 1998)
IEC 60404-8-3: 1985	Magnetic materials. Part 8: Specifications for individual materials. Section Three - Specification for cold-rolled magnetic non-alloyed steel strip delivered in the semi-processed state. (A new version will be published in 1998)
IEC 60404-8-4: 1986	Magnetic materials. Part 8: Specifications for individual materials. Section Four - Specification for cold-rolled non-oriented magnetic steel sheet and strip. (A new version will be published in 1998)
IEC 60404-8-5: 1989	Magnetic materials. Part 8: Specifications for individual materials. Section Five - Specification for steel sheet and strip with specified mechanical properties and magnetic permeability.
IEC 60404-8-6: 1986	Magnetic materials. Part 8: Specifications for individual materials. Section Six - Soft magnetic metallic materials. Amendment 1: 1992, (A new version will be published in 1999)

IEC Standards for Metallic Materials (Continued)

IEC 60404-8-7: 1988	Magnetic materials. Part 8: Specifications for individual materials. Section Seven - Specification for grain-oriented magnetic steel sheet and strip. Amendment 1: 1991, (A new version will be published in 1998).
IEC 60404-8-8: 1991	Magnetic materials. Part 8: Specifications for individual materials. Section Eight - Specification for thin magnetic steel strip for use at medium frequencies.
IEC 60404-8-9: 1994	Magnetic materials. Part 8: Specifications for individual materials. Section Nine - Standard specification for sintered soft magnetic materials.
IEC 60404-8-10: 1994	Magnetic materials. Part 8: Specifications for individual materials. Section Ten - Specification for magnetic materials (iron and steel) for use in relays.
IEC 60404-9: 1987	Magnetic materials. Part 9: Methods of determination of the geometrical characteristics of magnetic steel sheet and strip.
IEC 60404-10: 1988	Magnetic materials. Part 10: Methods of measurement of magnetic properties of magnetic sheet and strip at medium frequencies.
IEC 60404-11: 1991	Magnetic materials. Part 11: Method of test for the determination of surface insulation resistance of magnetic sheet and strip.
IEC 60404-12: 1992	Magnetic materials. Part 11: Guide to methods of assessment of temperature capability of interlaminar insulation coatings.
IEC 60404-13: 1995	Magnetic materials. Part 12: Methods of measurement of density, resistivity and stacking factor of electrical steel sheet and strip.

ASTM Standards for Magnetic Materials

Alternating Current

Test Methods for:

A 346 – 74 (1993)	Alternating Current Magnetic Performance of Laminated Core Specimens Using the Dieterly Bridge Method
A 932 – 95	Alternating-Current Magnetic Properties of Amorphous Materials at Power Frequencies Using Wattmeter-Ammeter-Voltmeter Method with Sheet Specimens
A 927/A 927M – 94	Alternating-Current Magnetic Properties of Torodial Core Specimens Using the Volmeter-Ammeter-Wattmeter Method
A 772/A 772M – 95	ac Magnetic Permeability of Materials Using Sine Current
A 912 – 93	Alternating Current Magnetic Properties of Amorphous Materials at Power Frequencies Using Wattmeter-Ammeter-Voltmeter Method With Torodial Specimens
A 697 – 91 (1996)	Alternating Current Magnetic Properties of Laminated Core Specimen Using the Voltmeter-Ammeter-Wattmeter Methods
A 804/A 804M – 94	Alternating-Current Magnetic Properties at Power Frequencies Using Sheet-Type Test Specimens
A 889/A 889M – 93	Alternating-Current Magnetic Properties at Low Inductions Using Wattmeter-Varmeter-Ammeter-Voltmeter Method and 25-cm Epstein Frame
A 343 – 93a	Alternating Current Magnetic Properties of Materials at Power Frequencies Using Wattmeter-Ammeter-Voltmeter Method and 25-cm Epstein Test Frame
A 348 – 95a	Alternating-Current Magnetic Properties of Materials Using the Wattmeter-Ammeter-Voltmeter Method, 100 to 10 000 Hz and 25-cm Epstein Frame

Amorphous Materials

Specification for:

A 901 – 90	Amorphous Magnetic Core Alloys Semi-Processed Types

Direct Current

Test Methods for:

A 341/A 341M – 95	Direct-Current Magnetic Properties of Materials Using D-C Permeameters and the Ballistic Test Methods
A 773/A 773M – 96	dc Magnetic Properties of Materials Using Ring and Permeameter Procedures with dc Electronic Hysteresigraphs
A 596/A 596M – 95	Direct-Current Magnetic Properties of Materials Using the Ballistic Method and Ring Specimens

ASTM Standards for Magnetic Materials (Continued)

Electrical Steel

Specifications for:

A 726 – 92	Cold-Rolled Magnetic Lamination Quality Steel, Semiprocessed Types
A 726M – 92	Cold-Rolled Magnetic Lamination Quality Steel, Semiprocessed Types (Metric)
A 345 – 90 (1995)	Flat-Rolled Electrical Steels for Magnetic Applications
A 876 – 92	Flat-Rolled, Grain-Oriented, Silicon-Iron, Electrical Steel, Fully Processed Types
A 876M – 92	Flat-Rolled, Grain-Oriented, Silicon-Iron, Electrical Steel, Fully Processed Types (Metric)
A 840 – 91 (1996)	Magnetic Lamination Steel, Fully Processed
A 840M – 91 (1996)'	Magnetic Lamination Steel, Fully Processed (Metric)
A 677 – 96	Nonoriented Electrical Steel, Fully Processed Types
A 677M – 96	Nonoriented Electrical Steel, Fully Processed Types (Metric)
A 683 – 91	Nonoriented Electrical Steel, Semiprocessed Types
A 683M – 91	Nonoriented Electrical Steel, Semiprocessed Types (Metric)

Test Methods for:

A 720 – 91	Ductility of Nonoriented Electrical Sheet Steel
A 721 – 92	Ductility of Oriented Electrical Sheet Steel
A 712 – 75 (1991)	Electrical Resistivity of Soft Magnetic Alloys
A 937 – 95	Interlaminar Resistance of Insulating Coatings Using Two Adjacent Test Surfaces, Determining
A 900 – 91 (1996)'	Lamination Factor of Amorphous Magnetic Strip
A 718 – 75 (1991)	Surface Insulation Resistivity of Multi-Strip Specimens (*Discontinued 1996†*)
A 717/A 717M – 95	Surface Insulation Resistivity of Single-Strip Specimens

Practice for:

A 664 – 93	Identification of Standard Electrical- and Laminations-Steel Grades in ASTM Specifications

Magnetic Amplifier Cores

Test Method for:

A 598 – 92	Magnetic Properties of Magnetic Amplifier Cores

Magnetic Shields

Test Method for:

A 698/A 698M – 92	Magnetic Shield Efficiency in Attenuating Alternating Magnetic Fields

Metallic Materials

Specifications for:

A 838 – 90a	Free-Machining Ferritic Stainless Soft Magnetic Alloys for Relay Applications
A 848/A 848M – 96	Low-Carbon Magnetic Iron
A 801/A 801M – 92	Iron-Cobalt High Magnetic Saturation Alloys
A 867/A 867M – 94	Iron-Silicon Relay Steels
A 753 – 85 (1990)	Nickel-Iron Soft Magnetic Alloys

Test Methods for:

A 900 – 91	Lamination Factor of Amorphous Magnetic Strip
A 719 – 90	Lamination Factor of Magnetic Materials
A 342 – 95	Permeability of Feebly Magnetic Materials

Practice for:

A 34 – 96	Sampling and Procurement Testing of Magnetic Materials

Nonmetallic Materials

Test Methods for:

A 893 – 86 (1991)	Complex Dielectric Constant of Nonmetallic Magnetic Materials at Microwave Frequencies
A 883 – 96	Ferrimagnetic Resonance Linewidth and Gyromagnetic Ratio of Nonmetallic Magnetic Materials
A 894/A 894M – 95	Saturation Magnetization or Induction of Nonmetallic Magnetic Materials

Sintered Powder Materials

Specifications for:

A 839/A 839M – 96	Iron-Phosphorous Powder Metallurgy (P/M) Parts for Soft Magnetic Applications
A 811 – 90	Soft Magnetic Iron Fabricated by Powder Metallurgy Techniques
A 904 – 90	50 Nickel - 50 Iron Powder Metallurgy Soft Magnetic Alloys

Terminology

Terminology Relating to:

A 340 – 96	Magnetic Testing

Bibliography

Bozorth, R. M. 1951. *Ferromagnetism*. Princeton, N. J.: D. Van Nostrand Co., Ltd.

Bradley, F. N. 1971. *Materials for Magnetic Functions*. New York: Hayden Book Co., Inc.

Brailsford, F. 1951. *Magnetic Materials*. London: Methuen and Co., Ltd.

Brailsford, F. 1966. *Physical Principles of Magnetism*. London: D. Van Nostrand Co., Ltd.

Chikazumi, S. 1964. *Physics of Magnetism*. New York: John Wiley and Sons.

DeMaw, M. F. 1981. *Core Design and Application Handbook*. Englewood Cliffs, N.J.: Prentice–Hall, Inc.,

Goldman, A. 1988. "Magnetic Ceramics." In *Electronic Ceramics*, ed. Lionel M. Levinson, pp. 147–189. New York: Marcel Dekker, Inc.

Grossner, N. 1983. *Transformers for Electronic Circuits*. New York: McGraw–Hill Book Co.

Heck, C. 1974. *Magnetic Materials and Their Applications*. New York: Crane–Russack and Co.

Hoshino, Y., Iida, S., and Sugimoto, M. 1971. *Ferrites: Proceedings of the International Conference*. Tokyo: University of Tokyo Press.

Hnatek, E. R. 1974. *Design of Solid State Power Supplies*. New York: Van Nostrand Reinhold.

Kampczyk, V. W., and Roess, E. 1978. *Ferrite Cores (Ferritkerne)*. Berlin: Siemens AG.

Kittel, C. 1971. *Introduction to Solid State Physics*. New York: John Wiley and Sons.

McLyman, C. T. W. 1982. *Magnetic Core Selection for Transformers and Inductors*. New York: Marcel Dekker, Inc.

Magnetic Materials Producers' Association. 1989. *Soft Ferrites: A User's Guide*. Evanston, Ill.: Magnetic Materials Producers Association.

Magnetics, Division of Spang and Co. 1989. *Ferrite Cores: Catalog FC509*. Butler, Penn.: Magnetics, Division of Spang and Co.

Mallinson, J. C. 1989. *The Foundations of Magnetic Recording*. San Diego: Academic Press.

Morrish, A. H. 1965. *The Physical Principles of Magnetism*. New York: John Wiley and Sons.

Olsen, F. 1966. *Applied Magnetism*. Eindhoven: Philips Technical Library.

Onoda, G. Y., and Hench, L. L. 1978. *Ceramic Processes Before Firing*. New York: John Wiley and Sons.

Palmer, H., Davis, R. F., and Hare, T. M. 1977. *Processing of Crystalline Ceramics*. Material Science Research, Vol 2. New York: Plenum Press.

Parker, R. J., and Studders, R. J. 1962. *Permanent Magnets and Their Application*. New York: John Wiley and Sons.

Philips Electronics Co. 1986. *Components and Materials; Book C-5*. Eindhoven: Philips Electronics and Materials Division.

Polydorf, W. J. 1960. *High Frequency Magnetic Materials*. New York: John Wiley and Sons.

Quartly, C. J. *Square Loop Ferrite Circuitry*. London: Iliffe Books, Ltd.

Rado, G., and Suhl, H. 1963. *Magnetism*, Vol. III. New York: Academic Press.

Reed, J. S. 1988. *Introduction to the Principles of Ceramic Processing*. New York: John Wiley and Sons.

Richards, C. E., and Lynch, A. C. 1953. *Soft Magnetic Materials for Telecommunications*. London: Pergamon Press, Ltd.

Roddam, T. 1963. *Transistor Inverters and Converters*. New York: Van Nostrand Reinhold.

Schieber, M. M. 1967. *Experimental Magnetochemistry*. Amsterdam: North-Holland Publishing Co.

Siemens AG. 1990–1991. *Ferrites: Data Book*. Munich: Siemens AG.

Smit, J. and Wijn, H. P. J. 1959. *Ferrites*. New York: John Wiley and Sons.

Smith, S. 1985. *Magnetic Components, Design and Applications*. New York: Van Nostrand Reinhold.

Snelling, E. C. 1988. *Soft Ferrites, Properties and Applications*. 2nd ed. London: Butterworths.

Snelling, E. C., and Giles. 1983. *Ferrites for Inductors and Transformers*. Letchworth: Research Studies Press, Ltd.

Soohoo, R. 1960. *Theory and Application of Ferrites*. Englewood Cliffs: Prentice–Hall, Inc.

Srivastava, M., and Patni, M. J. 1989. *Advances in Ferrites: Proceedings of the Fifth International Conference*, Vol. 1. New Delhi: Oxford and IBH Publishing Co. PVT.

Srivastava, M., and Patni, M. J. 1989. *Advances in Ferrites: Proceedings of the Fifth International Conference*, Vol. 2. Switzerland: Trans-Tech Publications.

Standley, K. J. 1962. *Oxide Magnetic Materials*. London: Oxford University Press.

TDK Corporation. 1987. *TDK Ferrite Cores for Power Supplies and EMI/RFI Filter*. Tokyo: TDK Corporation.

Tebble, R. S., and Craik, D. J. 1969. *Magnetic Materials*. London: Wiley Interscience.

Thompson, J. E. 1968. *The Magnetic Properties of Materials*. Cleveland: CRC Press.

Thomson LCC. 1988. *Soft Ferrites*. Book 1. Courbevoie: Thomson LCC.

Von Aulock, W. H. 1965. *Handbook of Microwave Ferrites*. New York: Academic Press.

Wang, F. F. Y. 1985. *Advances in Ceramics*. Vol. 15: Fourth International Conference on Ferrites—Part 1. Columbus, Ohio: American Ceramic Society.

Wang, F. F. Y. 1985. *Advances in Ceramics*. Vol. 16: Fourth International Conference on Ferrites—Part 2. Columbus, Ohio: American Ceramic Society.

Watanabe, H., Iida, S., and Sugimoto, M. 1981. *Ferrites, Proceedings of the ICF3*. Tokyo: Center for Academic Publications.

Watson, J. K. 1980. *Applications of Magnetism*. New York: John Wiley and Sons.

Wohlfarth, E. P. 1980. *Ferromagnetic Materials*, Vol. 2. Amsterdam: North-Holland Publishing Co.

Wood, P. 1981. *Switching Power Converters*. New York: Van Nostrand Reinhold.

APPENDIX 1
ABBREVIATIONS AND SYMBOLS

A	Cross-sectional area, cm^2
A	Attenuation or insertion loss, *dB*
A	Atomic weight
A	Angstrom units (10^{-8} m)
A_L	Inductance factor (inductance per 1000 turns), *mH*
A_c	Area of core, cm^2
A_W	Area of wire, cm^2 or circ. mils
A_p	Area product, cm^4
A_e	Effective area, cm^2
A_g	Cross-sectional area of gap
A_m	Cross-sectional area of magnet
a_0	Unit cell length, A
a	Distance between atoms, A
a	Anomalous loss coefficient
AC	Alternating current
B	Magnetic induction, gausses or teslas ($webers/m^2$)
B_s	Saturation induction
B_r	Remanent induction
B_m	Maximum induction or induction in magnet
B_g	Induction in gap
ΔB	Change in induction
B_e	Effective flux density
C	Centigrade temperature
C	Capacitance (farads)
C	Curie constant
C	Current-carrying capacity
c	Speed of light (3×10^{10} cm/s)
DA	Disaccommodation
DF	Disaccommodation factor
DC	Direct current

d	Thickness, cm
d	Density (g/cm^3)
d_o	Outer diameter
d_i	Inner diamter
d	Grain diameter, microns
E	Voltage, volts
E	Energy, ergs
E_p	Magnetostatic energy, ergs/cm^3
E_k	Anisotropy energy, ergs/cm^3
E_w	Wall energy, ergs/cm^2
E_l	Energy stored in an inductor, ergs/cm^3
E_{out}	Output voltage
E_{in}	Input voltage
e	Charge on an electron, coulombs
e	Efficiency
e	Eddy current coefficient (Legg)
F	Force, dynes
F	Farads (capacitance)
F_W	Winding factor
F_{cu}	Copper factor
f	Frequency, hertz
f_r	Resonant frequency, Hz
Δf	Bandwidth, Hz
f and F	Leakage factors, magnets
G	Gravitational constant
g	Spectroscopic splitting factor
H	Magnetic field strength, oersted or A/m
H_d	Demagnetizing field
H_m	Magnetic field in magnet
H_g	Magnetic field in gap
H_c	Coercive force, oersteds or A/m
H_{ci}	Intrinsic coercive force
Hz	Hertz, frequency
ΔH	Linewidth, oersted
ΔH_k	Spinwave linewidth, oersted
h	Planck's Constant
h_{crit}	Critical current (microwave power)
h	Hysteresis coefficient (Legg)
I	Current, amperes
I_p	Pulse current
J	Current density, A/cm^2
j	Unit imaginary vector

APPENDIX 1
ABBREVIATIONS AND SYMBOLS

K	Absolute temperature, Kelvin
K	Constant
K	Copper-fill factor
K_1, K_2	First and second anisotropy constants
K_W	Winding utilization factor (McLyman)
K_j	Current density factor
K_g	Geometrical factor
K_e	Electrical factor
k	Boltzman constant
L	Torque
L	Inductance, henries
L_N	Inductance for N turns
L_p	Parallel inductance
L_p	Pulse inductance
L_S	Series inductance
LF	Loss factor
l	Distance between poles
l_m	Magnetic path length, cm
l_e	Effective path length, cm
l_g	Length of gap
M	Intensity of magnetization, emu/cm^3
M_s	Saturation magnetization
MGO	Mega gauss-oersteds
m	Pole strength
m	Mass of an electron
m	Exponent of frequency in modified Steinmetz equation
N	Number of turns
N	Demagnetizing factor
N_p	Turns in primary winding
N_s	Turns in secondary winding
n	Number of unpaired electrons in an atom or ion
n	Exponent of B in modified Steinmetz equation
Oe	Oersteds, field strength
Pa	Pascals
P	Power, watts
P_O	Output power
P_i	Input power
P_t	Apparent power
P_t	Total loss, watts
P_c	Core loss, watts
P_w	Winding loss, watts
P_{hyst}	Hysteresis losses

P_e	Eddy current losses
P_r	Residual losses
p_{O_2}	Oxygen partial pressure
p	Total angular momentum of an electron
p	Porosity
Q	Quality factor—in a series inductance, $Q = X_{L/Rs}$
q	Electric charge in coulombs
R	Resistance, ohms
R_s	Series resistance
R_p	Parallel resistance
R	Reluctance, magnets
R_{th}	Thermal resistance
r	Distance between masses or charges
s	Skin depth
T	Temperature, degrees Kelvin or centigrade
T_C	Curie temperature
T_N	Néel temperature
TC	Temperature coefficient
TF	Temperature factor
T	Pulse width
t	Time, seconds
t_1, t_2	Specific times
t_{on}	On time of a transistor
t_{off}	Off time of a transistor
V	Volume, cm^3
V_m	Volume of a magnet
V_e	Effective volume
W_a	Window area, cm^3
X_L	Inductive reactance, ohms
X_C	Capacitive reactance, ohms
Z	Impedance, ohms
α	Regulation, %
$\alpha_1, \alpha_2 \ldots$	Direction cosines
δ	Angle between H and M
γ	Gyromagnetic ratio
δ	Thickness of domain wall
δ	Oxygen parameter (Tanaka)
ε	Phase angle between the direction of magnetization on the surface and at a depth in material
Δ	Change in a unit
η	Efficiency

APPENDIX 1
ABBREVIATIONS AND SYMBOLS

θ	Angle in degrees
θ	Angle between a magnetic field and the axis of a magnet
θ	Angle between the magnetization and the easy direction of magnetization
θ	Temperature rise in a power material, degrees
θ_C	Curie temperature
θ_N	Neel temperature
λ	Magnetostriction constant
λ_S	Saturation magnetostriction
λ_{100}	Magnetostriction in a cube edge direction
λ_{111}	Magnetostriction in a cube body diagonal direction
λ	Wavelength, meters
μ	Microns, 10^{-6} m
μ	Magnetic moment
μ_B	Bohr magneton
μ_0	Permeability of free space
μ	Permeability
μ_0	Initial permeability
μ_e	Effective permeability
μ_a	Amplitude permeability
μ_p	Pulse permeability
μ_{DC}	DC permeability
μ'	Real permeability
μ''	Imaginary part of permeability
μ^+	Clockwise rotating part of permeability of a circularly polarized electronmagnetic wave
μ^-	Counterclockwise part of permeability of above
μ_{max}	Maximum permeability
ρ	Resistivity, ohm-cm
σ	Applied stress
σ	Magnetic moment per unit weight, emu/g
τ	Pulse width, seconds
ϕ	Magnetic flux, maxwells or webers
χ	Magnetic susceptibility
ω	Angular velocity, radians/s

APPENDIX 2
ADDRESSES- MAJOR SUPPLIERS OF POWER ELECTRONICS COMPONENTS

Material Code
F = Ferrite Material
S = Metal Strip Material
A = Amorphous Metal Material
P = Metal Powder Core Material
N = Nanocrystalline Material

NORTH AMERICA
UNITED STATES

Allegheny-Ludlum Steel Corp. 500 Six PPG Place Pittsburgh, Pennsylvania, 15222	S
The Arnold Engineering Co. 300 N. West St. Marengo, IL 60152	S, A, P
AVX, A Kyocera Group Myrtle Beach SC	F
Epcos, Inc. 186 Wood Ave. S. Iselin, NJ 08830	F
Ceramic Magnetics 16 Law Drive Fairfield, NJ 07004	F
Fair-Rite Products Corp. P.O. Box J, One Commercial Row Wallkill, NY 12589	F
FerriShield 350 Fifth Ave. #7310 New York, NY 10118	F, S, P
Ferronics, Inc. 45 O'Connor Road Fairport, NY 14450	F

Hitachi Metals, America 2400 Westchester Ave. Purchase, NY 10577	S,A,F
Iskra Electronics 155 Dupont St. Plainview NY 11803	F
Metglas Division Honeywell Inc. 101 Columbia Rd. P.O. Box 1057 Morristown, NJ 07962	A
Magnetec GmbH Industriestrasse 7 D-63505 Langensebold, Germany	N
Magnetics, Div. of Spang & Co. P.O. Box 391 Butler PA 16003-0391	S,A,F,P
Micrometals 5615 E. LaPalma Ave. Anaheim, CA 92807	P
National Magnetics Group Inc 1210 Win Dr. Bethlehem PA 18017	F
MMG North America 126 Pennsylvania Ave. Patterson, NJ 07503-2512	F
Philips Ferrite Components Roswell, GA	F
Steward 1200 E. 36th Street P.O. Box 510 Chattanooga, TN 37401-0510	F
TDK of America 1600 Feehanville Dr. Mt. Pleasant, IL 60056	F,A

Tokin America 155 Nicholson Lane San Jose, CA 95134	F
TSC Ferrite International TSC Arnold Technologies 39105 N. Magnetics Blvd. P.O. Box 399 Wadsworth IL 60083	F S
TSC Pyroferric 507 Madison St. Toledo, IL 62489-0159	P
Coremaster International, Inc 180 South 300 West #120 Salt Lake City, Utah 84101	A, N.

Mexico

TDK de Mexico S.A. de C.V. Carr, Juarez Porvenir Parque Ind. AJ Bermudez, iCD Juarez, Chin. Mexico	F

Canada

Neosid, Canada, Ltd. 10 Vansco Rd. Toronto Ont. M8Z SJ4, Canada	F, P

South America

TDK DO Brasil Ind. E. Com Ltda Alameda Campinas 433 Conj. 111/112 Jardim Paulista CEP 01404-000 Sao Paulo, SP, Brazil	F

Asian Countries

Korea

TDK Korea 5th Fl. Sindo Bldg. 943-27 Daechi-dong Kangnam-ku Seoul, Republic of Korea	F
Samwha Electronics Co. Ltd. 142 Nonhyun-dong, Gangnam-ku Seoul, Korea	F
Isu Ceramics Co. C.P.O. Box 5680 Seoul, Korea	F
Korea Ferrite Co. 301, Dongjin Bldg. 218 2-ka Hankang-ro,Yongsan-ku Seoul, Korea	F
India Cosmo Ferrites Ltd 30Community Centre, Saket New Delhi, 11017, India	F
Hilversum Electronics Murugappa Electronics 29 II St. , Kamaraj Ave. Adyar, Madra, India	F
DGP Hinoday Industries Ltd. Bhosai Industrial Est. Pune,411026, India	F
International Ferrites Ltd.(an Epcos Co.) LB2, Sector III Salt Lake, Calcutta , 700091. India	F
Neosid (India) PVT Ltd. 144 Seevaram, Thoralpakkam Chennai 600-096, India	F, P

Japan

FDK Corp. 6-1-11 Shinbashi,Minato-ku Tokyo, 105 Japan	F

Tokin Corp. F,P
Hazama Bldg.
5-8 Kita-Aoyama 2-chome
Minato-ku
Tokyo, 107, Japan

TDK Corp. F,A
13-1 Nihonbashi 1-chome
Tokyo 103-8272, Japan

Mitsubishi Electric Corp. F
2-2 Marunouchi 2-chome
Chiyoda-ku
Tokyo, Japan

Hitachi Metals Co. F,A
2-1-2 Maunouchi, Chiyoda-ku
Tokyo, Japan

Murata Mfg. Co.,Ltd. F
26-10 Tenjin 2-chome
Nagaoka,kyo-shi
Kyoto,617, Japan

Nippon Ferrite Co F
1-25-1 Hyakunincho,Shinjuku-ku
Tokyo, 160, Japan

Sony Corp. F
7-35 Kitashinagawa 6-chome
Tokyo,11, Japan

Sumitomo Special Metals Co. S, F
22, Kitahama 5-chome, Higashi-ku
Osaka,541, Japan

Taiyo Yuden F
2-12 Ueno 1-chome Taitoh-ku
Tokyo,112, Japan

Tomita Electric Ltd. F
123 Saiwai Tottori-shi
Tottori-ken 680, Japan

Peoples Republic of China

Southwest Institute of Applied Magnetics F
P.O. Box 105, Mianyang
Sichuan China

Zhejiang Tiantong Electronics Co. Ltd F
11 Jianshe Rd. Guodian Town, Haining
Zhejiang, China

Hangzhou Linan Citong F
Magnetic Components Co.
3 Block, Government Administration Bureau
2# , Baoshi 1st Road, Hanzhou
Zheijiang, China

Hebei LC Electric Co. Ltd. F
No.8, Yucai, Laishui
Hubei, 074100, China

Yuxiang Magnetic Materials Inc. F
16F1 Jinyuan Bldg. No.57 Hubin S. Road
Xiamen, China

Jiangmen City Powder Metallurgy Factory Ltd. F
8 Longwan Road, Jiangmen
Guangdong, China

Singapore

Taiyo Yuden (Singapore) PTE, Ltd. F
9Joo Koon Rd.
Jurong Town
Singapore

Taiwan

Philips Electronic Industries Ltd. F
San Ming Bldg., 4th Floor
57-1 Chung Shan N. Rd.Sec.2
Taipei, Taiwan

Super Electronics Co., Ltd F

726, Chung Shan N. Rd.
Taipei Taiwan
TDK Electronics (Taiwan) Corp. F,A
159 Sec. 1, Chung Shan Rd.
Tatung Li, Yangmei
Taoyuan,Taiwan

Yeng Tat Electronics Co. Ltd F
P.O. Box 2-30 Shu-lin(238)
Taipei Hsien, Taiwan

Europe

United Kingdom

Salford Electrical Instruments F, P
Times Mills Heywood
Lancashire OL10 4N#, U.K.

Almag Ltd. F, P
17 Broomhills Rayne Rd.
Braintree, Essex, CM7 2Rg U.K.

Neosid Ltd. F, P
Edward House, Brownfields
Welwyn Garden City
Herfordshire, AL7 1AN, U.K.

France

AVX-Thomson50 Rue J-P Timbaud F
BP 13/92403
Courbevoie, Cedex, France

Czechoslavakia

Pramet F
Unecowska 2
787 53, Sumperk, Czechoslavakia

Germany

Kaschke KG GmbH and Co F

P.O. Box 2542
D-37015 Gottingen, Germany

Epcos AG Postfach 80 17 09 81617 Munich Germany	F
Vacuumschmelze GmbH Gruner Weg 37 P.O. Box 2253 D-6450 Hanau 1, Germany	S,A, N
Vogt Electronic AG D-94130, Obernzell, Germany	F, C

Netherlands

N.V. Philips Gloeilampenfabriken Commercial Department Materials Bldg. BE-4 Eindhoven, The Netherlands	F, P

Poland

Polfer Zaklad Materialow Magtczynch Warsaw-47 Poland	F

Yugoslavia

Iskra Feriti Stegne 29 61000 Ljubljana, Slovenia	F

APPENDIX 3
UNITS CONVERSION FROM CGS TO MKS (SI) SYSTEM

Symbol	Quantity	To convert from CGS Units	MKS Units	Multiply by Factor
l	Length	cm	m	10^{-2}
m	Mass	g	Kg	10^{-3}
F	Force	dyne	N, Newton	10^{-5}
E or W	Energy, Work	erg	Joule	10^{-7}
H	Magnetic Field Strength	Oersteds	A/m	79.58
B	Magnetic Induction or Flux Density	Gausses (Maxwell/cm^2)	Teslas(Wb/m^2) (volt-s)	10^{-4}
Φ	Magnetic Flux	Maxwells	Webers(Wb)	10^{-8}
μ	Permeability	Unitless	Henries/m	4×10^{-7}
F	Magnetomotive Force	Gilberts	Amp-turns	.7958
$(BH)_{max}$	Maximum Energy Product	Gauss-Oersteds (ergs)	Joules/m^3	7.96×10^{-3}

Other quantities which are the same in both systems and their common units include: Power, P (Watts, W); Charge, q (Coulombs); Potential, E or V (Volts); Time (Seconds); Resistance, R (ohms); Capacitance, C (Farads); Current, I (Amperes); Inductance, L (Henries).

Subject Index

Suppliers Index

Ferrites

Metal Strip

Powder Cores